T0299714

Metal Matrix Composites

With a focus on advances in metal matrix composite (MMC) fabrications from a theoretical and experimental perspective, this book describes the recent developments in the manufacturing of MMCs, various processing methods and parameters, mechanical properties and synthesis of MMCs. It deals with several multi-criteria decision-making techniques suggested to choose the best materials for application and the effects of reinforcement on chip formation, tool wear and part quality during the machining.

Features:

- Discusses modeling of MMCs and fabrication of hybrid MMCs
- Covers advanced characterization studies of nanocomposites
- Reviews high-temperature applications and cobalt-nickel combination materials
- Provides inputs regarding optimal selection of percentage of reinforcement materials for MMC's fabrication based on industrial requirements
- Focuses on aerospace and automotive industries

This book is aimed at graduate students, researchers and professionals in micro/nanoscience and technology, mechanical engineering, industrial engineering, metallurgy and composites.

Metal Matrix Composites
Advances in Processing Methods, Machinability Studies and Applications

Edited by
Jayakrishna Kandasamy, Rajyalakshmi G and
Mohamed Thariq Hameed Sultan

CRC Press
Taylor & Francis Group
Boca Raton London New York

CRC Press is an imprint of the
Taylor & Francis Group, an **informa** business

Designed cover image: © Shutterstock

First edition published 2023
by CRC Press
6000 Broken Sound Parkway NW, Suite 300, Boca Raton, FL 33487-2742

and by CRC Press
4 Park Square, Milton Park, Abingdon, Oxon, OX14 4RN

CRC Press is an imprint of Taylor & Francis Group, LLC

Library of Congress Cataloging-in-Publication Data
Names: Kandasamy, Jayakrishna, editor.
Title: Metal matrix composites : advances in processing methods, machinability studies and applications / edited by Jayakrishna K., Rajyalakshmi G., M.T.H. Sultan.
Other titles: Metal matrix composites (CRC Press)
Identifiers: LCCN 2022041642 (print) | LCCN 2022041643 (ebook)
Subjects: LCSH: Metallic composites.
Classification: LCC TA481 .M4453 2023 (print) | LCC TA481 (ebook) |
DDC 620.1/6–dc23/eng/20221122
LC record available at https://lccn.loc.gov/2022041642
LC ebook record available at https://lccn.loc.gov/2022041643

ISBN: 9781032385235 (hbk)
ISBN: 9781032385259 (pbk)
ISBN: 9781003345466 (ebk)

DOI: 10.1201/9781003345466

Typeset in Times
by codeMantra

Contents

Editors

Jayakrishna Kandasamy is an Associate Professor in the School of Mechanical Engineering at the Vellore Institute of Technology University, India. His research is focused on bio-composites for biomedical and ballistic applications. He has published 66 journal publications in leading SCI/SCOPUS Indexed journals, 26 book chapters, 91 refereed conference proceedings and seven books in CRC/Springer Series. He has been awarded Global Engineering Education Award from Industrial Engineering and Operations Management (IEOM) Society International, U.S.A. in 2021, Institution of Engineers (India) – Young Engineer Award in 2019.

Rajiyalakshmi G is an Associate Professor in the School of Mechanical Engineering at the Vellore Institute of Technology University, India. Her research is focused on advanced materials, laser peening, advanced machining, mathematical modeling, sustainable manufacturing and optimization. She has published 51 journal publications in leading SCI/SCOPUS Indexed journals, nine book chapters and 20 refereed conference proceedings.

Mohamed Thariq Hameed Sultan completed his PhD in 2011 from the Department of Mechanical Engineering, University of Sheffield, United Kingdom in the field of Mechanical Engineering. He specializes in the fields of Hybrid Composites, Advance Materials, Structural Health Monitoring and Impact Studies. He is also a Professional Engineer (PEng) registered under the Board of Engineers Malaysia (BEM), and he is also a Charted Engineer (CEng) registered with the Institution of Mechanical Engineers (IMechE) United Kingdom. Recently, he was also awarded the Professional Technologist (PTech) from Malaysian Board of Technologist (MBOT). He has published more than 291 journal articles and also published almost 25 books internationally.

Preface

With the ever-increasing consumer demand for systems and machines that are more energy efficient, stronger, lightweight and cost effective, the search for new and advanced materials remains a subject of interest all time. Materials and their properties are to be explored for continuous improvement in operational and industrial standards, with appraising technological developments and composites playing a vital role in achieving this. Especially, metal matrix composites (MMCs) are a class of materials that have proven successful in meeting most of the rigorous specifications in applications where lightweight, high stiffness and moderate strength are the requisite properties. The variety of reinforcement materials and flexibility in their primary processing offer great potential for the development of composites with the desired properties for unique applications.

This book focuses on the advances in processing and development of MMCs for marine, novel, aerospace and other industrial applications. In this book, the progress, utilization and future potential of MMCs in various industrial and commercial applications are discussed, together with the existing challenges hindering their full market penetration. New research on novel modelling strategies and/or simulation tools to establish the processing-structure-properties link in materials from atomic-to macroscale. This book focuses on addressing innovative theoretical or computational studies of material structure and behaviour, examining materials evolution at electronic-, atomic- and meso-scales, in relation to mechanical, chemical, electronic, optical, biological, or biomedical properties. It also focuses on machine learning, learning systems and data-driven tools for microstructural analysis, processing and properties simulation and materials discovery. It is intended to provide basic- and advanced-level knowledge to cater the needs of both beginners and specialists.

Jayakrishna Kandasamy
Rajyalakshmi G
Mohamed Thariq Hameed Sultan
MATLAB® is a registered trademark of The MathWorks,
Inc. For product information,
please contact:
The MathWorks, Inc.
3 Apple Hill Drive
Natick, MA 01760-2098 USA
Tel: 508-647-7000
Fax: 508-647-7001
info@mathworks.com
www.mathworks.com

Contributors

Abhishek Singh
School of Mechanical Engineering
Vellore Institute of Technology
    University
Vellore, Tamil Nadu, India

S. K. Ariful Rahaman
School of Mechanical Engineering
Vellore Institute of Technology
Vellore, Tamil Nadu, India

K. Balamurugan
Department of Mechanical Engineering
VFSTR (Deemed to be University)
Guntur, Andhra Pradesh, India

Ranjeet Kumar Bhagchandani
School of Mechanical Engineering
Vellore Institute of Technology
Vellore, Tamil Nadu, India

A. Chinnamahammad Bhasha
Department of Mechanical Engineering
VFSTR (Deemed to be University)
Guntur, Andhra Pradesh, India

A. Deepa
School of Mechanical Engineering
Vellore Institute of Technology
Vellore, Tamil Nadu, India

T. Deepthi
Department of Mechanical Engineering
VFSTR (Deemed to be University)
Guntur, Andhra Pradesh, India

P. Dilip Kumar
School of Mechanical Engineering
Vellore Institute of Technology
    University
Vellore, Tamil Nadu, India

Dileep Kumar Ganji
School of Mechanical Engineering
Vellore Institute of Technology
Vellore, Tamil Nadu, India

Y. Jyothi
Faculty of Mechanical Engineering
VFSTR (Deemed to be University)
Guntur, Andhra Pradesh, India

Jayakrishna Kandasamy
School of Mechanical Engineering
Vellore Institute of Technology
Vellore, Tamil Nadu, India

Sajan Kapil
Department of Mechanical Engineering
Indian Institute of Technology
    Guwahati
Guwahati, Assam, India

Nobel Karmakar
Department of Mechanical Engineering
Indian Institute of Technology
    Guwahati
Guwahati, Assam, India

T. P. Latchoumi
Department of Computer Science and
    Engineering
SRM Institute of Science and
    Technology
Ramapuram, Tamil Nadu, India

Manas Mohan Mahapatra
School of Mechanical Sciences
Indian Institute of Technology
    Bhubaneswar
Bhubaneswar, Odisha, India

N.V.S.S.S.K. Manne Dilip
School of Mechanical Engineering
Vellore Institute of Technology
    University
Vellore, Tamil Nadu, India

Arabinda Meher
School of Mechanical Sciences
Indian Institute of Technology
Bhubaneswar, India

D. V. V. Pavan Kumar
CMR Institute of Technology
Ananthalakshmi Institute of Technology
    and Sciences
Hyderabad, Andhra Pradesh, India

G. Rajyalakshmi
School of Mechanical Engineering
Vellore Institute of Technology
Vellore, Tamil Nadu, India

M. Ramakrishna
Department of Mechanical
    Engineering
VFSTR (Deemed to be University)
Guntur, Andhra Pradesh, India

R. Ramanujam
School of Mechanical Engineering
Vellore Institute of Technology
Vellore, Tamil Nadu, India

G. Ranjith Kumar
Department of Mechanical
    Engineering
Sri Venkateswara College of
    Engineering and Technology
Chittoor, Andhra Pradesh, India

Priyaranjan Samal
School of Mechanical Sciences
Indian Institute of Technology
    Bhubaneswar
Bhubaneswar, Odisha, India

T. Sampath Kumar
School of Mechanical Engineering
Vellore Institute of Technology
    University
Vellore, Tamil Nadu, India

Ritam Sarma
Department of Mechanical Engineering
Indian Institute of Technology
    Guwahati
Guwahati, Assam, India

M. Sathishkumar
Department of Mechanical Engineering
Amrita School of Engineering, Amrita
    Vishwa Vidyapeetham
Chennai, Tamil Nadu, India

Shivaprasad Satla
Department of CSE
Malla Reddy Engineering College (A)
Secunderabad, India

M. Vignesh
Department of Mechanical Engineering
Amrita School of Engineering, Amrita
    Vishwa Vidyapeetham
Chennai, Tamil Nadu, India

S. Vigneshwaran
Faculty of Mechanical Engineering
Kalasalingam Academy of Research
    and Education
Srivilliputtur, Tamil Nadu, India

A. Vinoth Jebaraj
School of Mechanical Engineering
Vellore Institute of Technology
    University
Vellore, Tamil Nadu, India

T. Vishnu Vardhan
CMR Institute of Technology
Anantha Lakshmi Institute of
    Technology and Sciences
Hyderabad, Andhra Pradesh, India

Pandu R. Vundavilli
School of Mechanical Sciences
Indian Institute of Technology
    Bhubaneswar
Bhubaneswar, Odisha, India

Balram Yelamasetti
CMR Institute of Technology
Anantha Lakshmi Institute of
    Technology and Sciences
Hyderabad, Andhra Pradesh, India

1 Processing and Characterization of Aluminum Metal Matrix Composites by Stir Casting with Carbide Reinforcement

Priyaranjan Samal
Indian Institute of Technology Bhubaneswar

Pandu R. Vundavilli
Indian Institute of Technology Bhubaneswar

Arabinda Meher
Indian Institute of Technology Bhubaneswar

Manas Mohan Mahapatra
Indian Institute of Technology Bhubaneswar

CONTENTS

DOI: 10.1201/9781003345466-1

1.1   INTRODUCTION

The demand for metal matrix composites (MMCs) in automobile and aerospace structures is increasing in the past two decades for their excellent strength to weight ratio, high stiffness, high wear, corrosion resistance, etc. when compared to the monolithic materials [1]. MMCs can be defined as the combination of two constituents, where one is a metal or its alloy called a matrix, whereas the other is a non-metallic or organic compound termed reinforcements. The matrix materials are soft and ductile, whereas the reinforcement materials are preferably ceramic or organic materials, which are hard and brittle in nature. To the soft and ductile matrix material, a hard reinforcement phase is incorporated to produce enhanced properties than the base matrix. Among the existing materials for matrix element (Al, Mg, Cu, Fe) for production of MMCs, aluminum is the most commonly used for its less density, good machinability, and easiness in processing. Magnesium materials have a limitation as they are more reactive to higher temperatures with low fracture resistance. Although steel has excellent wear-resistant properties, it cannot be used in oxide and marine environments for a longer duration. Similarly, the low strength of copper restricts its usage in many structural applications. With the addition of reinforcements in aluminum and its alloys, the MMCs attribute in improving the mechanical and wear properties that become compatible with its applications. Further need for advanced materials under the category of aluminum MMCs is the primary focus of the researchers, which can yield light-weight composites with enhanced strength. The automobile industries prefer aluminum composites due to their less weight as compared to other materials [2].

Modification of properties can be achieved with the help of a wide range of reinforcement materials to aluminum and its alloys based on the applications. Although aluminum and its alloys provide moderate to high strength, the major hurdle in advanced engineering applications is their low resistance. Therefore, by adding carbide particles such as SiC, TiC, B_4C, WC, etc., the wear resistance, corrosion resistance, hardness, and strength can be amended for aluminum MMCs with low production cost [3–6]. In this chapter, carbide reinforcements are considered since these are easily dispersed in the aluminum matrix, and also readily available. The easiness in dispersion positively affected the interfacial bonding between the Al matrix and carbide reinforcement, which causes improvement in mechanical properties. Furthermore, carbide reinforcements do not react with aluminum at high temperatures. Table 1.1 gives some basic properties of selected carbide reinforcement material.

The processing methods play an important role in fabricating aluminum MMCs to comply with the present industrial needs. For expanding the uses of these aluminum MMCs, cost-effective methods are necessary for the production of composites. Among different molten metal processing methods, stir casting is considered the most reliable and cost-effective method [7,8]. Stir casting involves the introduction of reinforcement materials into the selected molten matrix by stirring action. The stirring action also helps in maintaining the reinforcement particles in a suspension state. Still, there are challenges associated with the stir casting process, i.e. uniform dispersion, wettability, less porosity, etc. The wettability refers to the interfacial

TABLE 1.1

Common Properties of Carbide Reinforcements [9]

Properties	SiC	TiC	WC	B4C
Density (g/cm³)	3.16	4.93	15.63	2.52
Melting point (°C)	1,955	3,067	2,870	2,763
Hardness, Vickers (GPa)	35–45	24–32	24–30	38
Modulus of elasticity (GPa)	448	400	710	460
Thermal conductivity (W/m K)	16–20	17–32	110	17–42
Crystal structure	Hexagonal	Cubic	Hexagonal	Rhombohedral

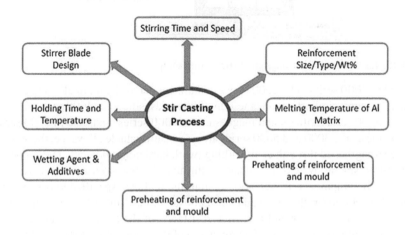

FIGURE 1.1   Parameters involved in the stir casting process.

bonding between the matrix and reinforcement. Good wettability indicates excellent interfacial strength, thus enhancing the mechanical strength of the composite. The non-homogeneous distribution also arises from the undesirable chemical reaction between the matrix and reinforcement particles. The clustering of reinforcement particles is also a major concern in achieving good mechanical properties. Addressing the correct stoichiometric calculation of the elements while adding in the composite with proper design the stir casting set up, various problems can be minimized during the fabrication. Figure 1.1 shows different parameters that affect the fabrication of Al MMCs in the stir casting process, whereas the schematic representation of the stir casting is shown in Figure 1.2.

1.2   GENERALLY CONSIDERED AL MATRIX MATERIAL

Depending upon the application, aluminum and its alloys are considered for the preparation of composite materials. Pure aluminum is rarely used in industrial applications nowadays for its inferior properties than its alloys. Aluminum alloys are classified into different series based on the principal alloying element. Pure aluminum is

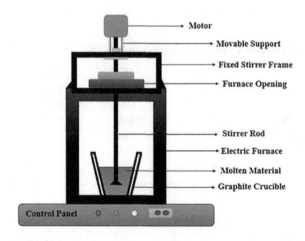

FIGURE 1.2 A schematic diagram of stir casting setup.

notified as 1000 series. The 2000 series contain copper as the main alloying element with aluminum; similarly, manganese in 3000 series, silicon in 4000 series, magnesium in 5000 series, silicon and magnesium in 6000 series, and zinc in 7000 series. Among these, the 3000 and 5000 series are non-heat-treatable alloys, i.e. these alloys can be strengthened through cold working (work hardening) process. The properties of heat-treatable alloys can be enhanced through the hot working (age hardening) process. The graphical representation of aluminum alloys' specifications is given in Figure 1.3. Moreover, the 6000 and 7000 alloys are best suited for structural application, whereas the wear and corrosion resistance of the 5000 series are excellent when compared to other alloys. A comparison of the strengths of different aluminum alloys is given in Figure 1.4.

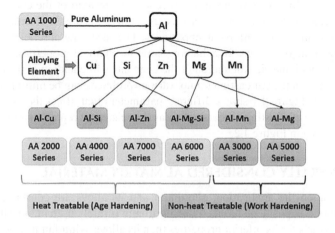

FIGURE 1.3 Classification of aluminum alloys.

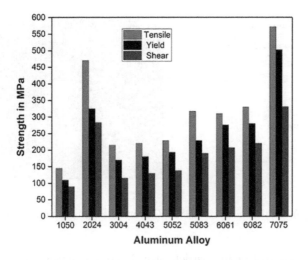

FIGURE 1.4 A comparison of the strength of various aluminum alloys (approx. values).

1.3 REINFORCEMENT PHASE

In the liquid metallurgy process, the mechanical properties of Al MMCs are mainly dependent upon the types of reinforcement phases. The liquid metallurgy, especially the stir casting process, totally favors any types of the reinforcement particles, i.e. metallic, ceramic, organic, and non-metallic. This section deals with the different carbide reinforcement particles used in Al MMCs for many industrial applications.

1.3.1 SILICON CARBIDE (SiC) AS REINFORCEMENT MATERIAL

Many researchers have worked in the field of microstructural characterization of aluminum MMCs reinforced with SiC. Stir casting method was adopted for the fabrication of Al-Si alloy reinforced with SiC particles, where a clean interface with homogenously distributed particles was observed throughout the composites [10]. The stir casting was aligned with the vortex free mechanism which enhanced the mechanical performance of the composites. The influence of nano and micro SiC on the microstructural and mechanical properties of A356 MMCs was investigated by stir casting [11]. The nano-sized SiC-reinforced composites attributed to the higher tensile strength when compared to the microparticles, as shown in Figure 1.5. The nanoparticle causes more interactions resulted from the dislocation and grain refinement, thus exhibiting higher strength of the composites.

Similarly, the influence of SiC particle size and content on the mechanical properties of the AA7075 composites was investigated through the stir casting route [12]. With an increase in the particulate size and higher weight percentage of SiC, the tensile strength and hardness of the MMCs significantly enhanced. This implies a better bonding between the Al alloy and SiC particles. This can also be referred to the fact that SiC restricts the dislocation movement and leads to better bonding of the composites. With an increase in the sliding distance, the wear rate of the MMCs

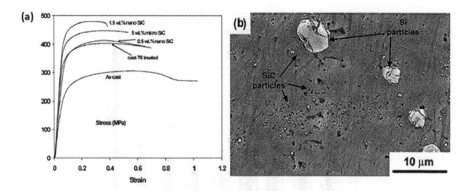

FIGURE 1.5   (a) Tensile strength of A356/SiC composites. (b) SEM microstructure [11].

is reduced irrespective of the particle size. Many researchers have reported the SiC-reinforced aluminum MMCs with a scope in the brake application in the automobile sector. Aluminum SiC fabricated through stir cast method utilized in making brake drum was analyzed in comparison with brake drum made up of cast iron [13]. The effect of speed and braking load on the coefficient of friction of the composites was investigated, whereas the interface temperature generated between the brake drum and shoe also affected the braking efficiency of the composites. The higher generated coefficient of friction of the heat-treated composites than the cast iron limited their application. Moreover, Kumar et al. [14] synthesized MMCs through stir casting used AA6061 and SiC reinforcement, which resulted in higher mechanical and wear properties. The ultimate tensile strength was enhanced with higher SiC content, whereas the hardness increased simultaneously with a reduction in ductility. This can be attributed to the fact that the hard SiC phase causes more stress concentration sites, which makes the composite vulnerable for crack initiation. The effect of nano SiC reinforcement on the corrosion and wear properties of the 6061 composites was reported [15]. The SiC particle in its nano form does not react with any corrosive solution or has very less impact. Therefore, the corrosive resistance of the aluminum composites was found higher as compared to the base alloy. The wear resistance was also seen to be higher when compared to the base alloy both in dry and corrosive atmospheres. SiC reinforced with aluminum alloy was reported to have excellent impact strength with their use in ballistic materials [16]. Among the aluminum alloys 2124, 5083, and 6063, the MMCs fabricated using 6063/SiC exhibited higher impact strength with the test conditions. The reduction in SiC resulted in the formation of crack nucleation sites which affected the impact strength of the composites. The fractography of the Al/SiC composites is shown in Figure 1.6. Aluminum silicon alloy reinforced with SiC particles was utilized for the analysis of corrosion behavior of the composites through stir casting [17]. Irrespective of the applied load, the wear rate of the composites was reduced with respect to the higher content of SiC. Moreover, the increase in the SiC percentage also enhanced the corrosion resistance of the MMCs up to 50% when compared to that of base alloy, providing scope of using these MMCs in marine and automobile applications. Singh et al. [18]

FIGURE 1.6 Fractography of 6063/SiC composites showing crack surfaces [16].

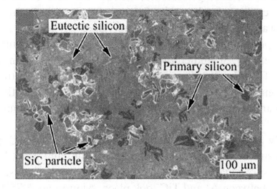

FIGURE 1.7 Micrograph showing good interface of SiC particles with Al alloy [18].

studied the mechanical and wear characteristics of the Al-Si alloy reinforced with SiC composites and reported a 38% increase in the tensile strength when compared with the base alloy. Moreover, with an increase in the applied load, the wear rate was found to be increased, but this was independent of the sliding distances. The micrograph of the interface of the Al/SiC composite is shown in Figure 1.7.

1.3.2 TITANIUM CARBIDE (TIC) AS REINFORCEMENT MATERIAL

Titanium carbide (TiC), being a hard ceramic phase, provides excellent hardness properties upon reinforcing with aluminum alloy. Samal et al. [4] synthesized aluminum 5052/TiC composites adopting the in-situ stir casting route. The in-situ method exhibited homogenous dispersion due to exothermic chemical reaction happening in between the aluminum and reinforced particles. The excellent wear resistance offered by the MMCs was found suitable for brake disc applications in the automobile sector. Due to the brittleness of the composites, the impact strength was found to be reduced, which indicates the loss of ductility. Similarly, the nano formation of

TiC particles in the aluminum composites strengthened the composite at ambient temperature [19]. The grain refinement and restriction of dislocation movement as a result of the uniform dispersion in the composites leads to the higher strength. Moreover, this phenomenon also restricts the chances of crack nucleation and its propagation. The TEM micrograph of the interface of the TiC particles is shown in Figure 1.8. The aluminum 6063 TiC composites were fabricated by stir casting method with a homogenous distribution of TiC particles using a mechanical stirrer [20]. The brittle nature of the composite favors the minor crack formation with an increase of stress concentration areas near the Al/TiC interface. The above phenomenon resulted in debonding with a subsequent reduction in the impact strength of the MMCs. However, the impact strength of the composites was reduced as a result of the loss of ductility while reinforcing TiC into aluminum matrix [21]. The enhanced stir casting method was adopted for the fabrication of AA6061/TiC composites where the wear resistance was found to be improved when compared to the base alloy [22]. With the increase in the sliding velocities, the rubbing surfaces were subjected to a higher temperature where the thermal softening of the matrix occurred. As a result, more wear loss was found with higher sliding velocity. The wear rate of the composite with more percentage of TiC was found to be reduced as compared to that of the alloy. Kumar et al. [23] studied the effect of addition of TiC on the mechanical characterization of Al-Cu alloy composites. The ultimate tensile strength and yield strength of the composites were increased by 19% and 12%, respectively, upon addition of titanium carbide. The dimple structure revealed the ductile nature of fracture in tensile testing.

Homogenously dispersed and equiaxed spherical-shaped TiC particles were observed as a result of the in-situ reaction with aluminum alloy [24]. Moreover, the absence of any residual secondary phases in the composites suggests that the complete reaction between Al and TiC took place. Aluminum 7075/TiC composites are considered for aerospace applications for their excellent mechanical and wear

FIGURE 1.8   TEM micrograph showing the Al TiC interface [19].

properties [25]. The Orowan strengthening mechanism favored the enhancement of tensile strength with grain refinement of the TiC particles. The composites also possess higher bending stress due to the hard nature of TiC reinforcement. The base alloy exhibited a lower value of the coefficient of friction as compared to the composites, whereas the wear resistance of the composites was found to be higher when compared to the base alloy. Stir casting by direct reaction synthesis method was adopted for the fabrication of AA7079/TiC composites [26]. The in-situ formed TiC particles were distributed uniformly in the center of the grain. The effect of the formation of the grain boundary with the TiC interface on its mechanical properties was extensively discussed in this work. This research also provides a deep insight into the formation of in-situ TiC particles in the aluminum phase. Sujith et al. [27] reported the abrasive wear behavior of the TiC-reinforced 7079 MMCs via stir casting process where good wettability resulted in the enhancement of wear resistance upon reinforcing TiC into aluminum alloy composites. With the higher content of TiC particles, the wear loss was found to be reduced, whereas the wear loss was linearly increased with the increase in the sliding distance during the wear experimentation.

1.3.3 Tungsten Carbide (WC) as Reinforcement Material

Tungsten carbide is a hard and rigid material with very high strength. The higher resistance to deformation with efficient tensile strength makes it suitable for the consideration of reinforcement in aluminum MMCs. Also, the toughness and excellent wear resistance properties give it an alternate combination for the tribological applications. Ravikumar et al. [6] synthesized AA6082/WC composites via stir casting and reported its mechanical behavior. The composites exhibited better interfacial bonding which resulted in higher tensile and yield strength with a higher content of tungsten carbide with a reduction in ductility. With higher hardness, the impact strength was found to be reduced when the wt% of reinforcement increased due to less plastic deformation energy. The variations of tensile strength and impact strength are shown in Figures 1.9 and 1.10, respectively.

Aluminum LM4 alloy composites reinforced with tungsten carbide particles were investigated by stir casting method [28]. The tensile strength and impact strength

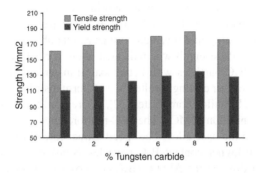

FIGURE 1.9 Variation of tensile strength with wt% of tungsten carbide [6].

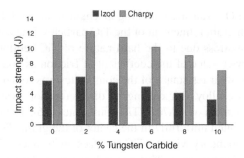

FIGURE 1.10 Variation of impact strength with wt% of tungsten carbide [6].

were observed to be enhanced with the addition of WC particles, whereas the incorporation of hard WC particles reduced the ductility of the MMCs. Moreover, with the higher content of WC in the composites, the wear rate was found to be decreased which shows higher wear resistance. The base alloy has undergone more plastic deformation during the wear test, whereas in the case of the WC reinforced composites, less plastic deformation was observed as shown the WC particles, which restricted the plastic flow. Dhas et al. [29] studied the mechanical behavior of the AA5052 hybrid composites considering WC and SiC particles both with graphite. The tungsten carbide-reinforced composites exhibited better tensile strength than SiC-reinforced composites showing excellent wettability.

1.3.4 BORON CARBIDE (B_4C) AS REINFORCEMENT MATERIAL

Boron carbide is one of the hardest materials (next to diamond and CBN) that finds many applications in the abrasive domain. Aluminum composites reinforced with boron carbide are also used in nuclear applications for shielding nuclear waste. Aluminum 6061/B_4C composites were fabricated through the stir casting method to study their mechanical and tribological properties [30]. The ultimate tensile strength was significantly increased with a higher content of B_4C. Moreover, the ductility was also seen to be improved in an appreciable quantity, showing that the composites retained their ductile behavior upon reinforcement of hard boron carbide materials. Further, the wear resistance of the composite was found to be enhanced when compared with the base alloy. Kalaiselvan et al. [31] have fabricated Al 6061 B_4C composites via stir casting and revealed similar results about tensile strength. The increment in the tensile strength was indicated about a clear interface bonding of the aluminum and B_4C particles with better load transfer carrying capacity of the composites. Moreover, stir casting resulted in a uniform distribution of boron carbide particles in the AA2014 alloy composites [32], whereas a little cluster was observed in the higher concentration MMCs. The B_4C-reinforced composites showed higher wear resistance than the base alloy. The specific wear rate of the MMCs was found to be reduced with higher content B_4C particles. Similarly, the effect of thermal and mechanical loading on the Al 6061/B_4C composites was investigated [33]. The

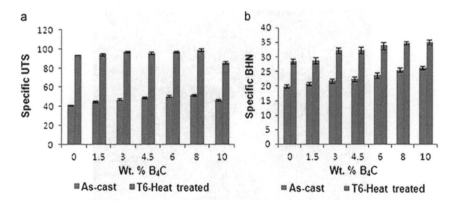

FIGURE 1.11 Variation of specific UTS and hardness with respect to B_4C content [33].

composites were subjected to stir casting followed by the extrusion process for the evaluation of their mechanical behavior. The heat treatment process enhanced the specific tensile strength and hardness of the as-cast specimen. The variations of specific ultimate tensile strength and specific hardness are shown in Figure 1.11. The boron carbide-reinforced Al composites were utilized as the brake application in automotive industries [34].

The impact strength of boron carbide-reinforced Al composites was reported by Ibrahim et al. [35]. The wettability was improved by the addition of B_4C particles in the composites. It was revealed that the formation of precipitation in the composites governs the impact toughness of MMCs. The case studies also reported that the elimination of debonding occurred by the presence of Zr and Sc as protecting layers. Toptan et al. [36] reported that the poor wetting due to the low temperature exhibited relatively lower mechanical properties even if the reinforcements were homogeneously distributed in the composites. Further, Pozdniakov et al. [37] investigated two composites considering Al 6063 and 1545 with B_4C as reinforcement for nuclear applications such as compact storage for utilized nuclear fuel. The purity in the boron carbide particles facilitated good interfacial bonding with aluminum alloy. Rolling as the secondary process eliminated agglomerations and improved distribution all over the composites. Kerti and Toptan [38] studied the microstructural behavior of Al/B_4C composites. The bigger particle size resulted in uniform distribution with minimum agglomeration when compared to smaller particles. The micrograph of the Al/B_4C composites is shown in Figure 1.12. Similarly, three aluminum composites were fabricated with B_4C, SiC, and Al_2O_3 through stir casting where it was reported that B_4C-reinforced MMCs have shown comparatively higher mechanical strength than the other two composites with better interfacial bonding [39]. Khan et al. [40] studied the creep behavior and hot hardness of Al-Mg alloy reinforced with B_4C particles by the stir casting process. With an increase in the B_4C content, it leads to restriction in the dislocation motion, and thus attributed to the creep activation energy.

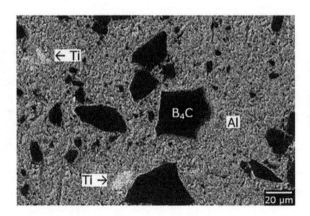

FIGURE 1.12 Micrographs of Al/B$_4$C composites [38].

1.4 APPLICATION OF AL MMCs

Aluminum MMCs are focused on further advancement in many fields for their low weight with high strength combinations. Aluminum MMCs exhibit a good combination of properties when compared to the conventional material in industrial fields like automobile, aviation, electronics, defense sector, etc. The following are a few important applications:

i. Automobile – Aluminum MMCs are preferred in many automobile applications especially in parts like brake disc, connecting rod, piston, etc. The low density favors aluminum with suitable reinforcement particles for these appliances. Al with SiC is the most preferred in automotive vehicles due to the excellent wear resistance with minimum noise generation, thereby increasing energy efficiency. With high thermal conductivity and good wear resistance, aluminum MMCs are substituted in engine parts like piston and connecting rod. The high toughness properties with good machinability favor to manufacture chassis parts.

ii. Marine, rail, and aircraft transports – Aluminum MMCs are extensively used in the manufacturing of rail cars for its less weight and anti-corrosion behavior. The corrosion resistance also comes effectively in the marine sector especially in Al 5083 and 5052 alloy composites. Fuel efficiency and speed are also increased by using light-weight Al MMCs in ship construction. Light weight with good stiffness and low thermal expansion coefficient makes Al MMCs suitable for aerospace applications. Nowadays aircraft primary wings, fuselage, fan blades in turbine engines, and propellers are exclusively replaced by Al MMCs.

iii. Electronics – Heat dissipation is a major issue in many of the electronic equipment. The low thermal expansion coefficient and high thermal conductivity behavior of Al MMCs overcome this issue of heat dissipation up to a large extent, especially with carbide reinforcements.

iv. Defense sector – High rigidity with low-cost properties has a great effect on replacing modern weapons. The toughness is a major criterion while designing any sophisticated instrument in the defense arena. Many kinds of researches are going on to replace the current artillery with aluminum MMCs.

1.5   CONCLUSION

The rapid growth in technological advancement requires light-weight materials as per the need of the current manufacturing application. Aluminum MMCs can fulfill the desired tribological and mechanical properties when reinforced with carbide constituents, especially silicon and titanium carbide particles. A lot of organic compounds and industrial wastes are also given the emphasis on the production and utilization to minimize the environmental issue.

- Aluminum MMCs are well deserved for mainly the automobile, aerospace, and defense sectors.
- Stir casting is the most economical and viable process for manufacturing of aluminum MMCs.
- Mechanical properties of the stir casting-processed aluminum MMCs mainly depend upon dispersion, size/shape, quantity of reinforcement, stirring time, and reaction time during the manufacturing process.
- In recent years, many types of researches come out in the fields of Al MMCs and achieved many goals in developing new materials to overcome the issues related to the performance of running materials. Moreover, the field of Al MMCs still needs attention and research in the areas related to fabrication and amalgamation of new reinforcement phase. There is always scope for research in the micro and nano constituents in Al MMCs and their behavior in achieving the required properties.

REFERENCES

1. N. K. Bhoi, H. Singh, and S. Pratap, "Developments in the aluminum metal matrix composites reinforced by micro/nano particles – A review," *J. Compos. Mater.*, vol. 54, no. 6, pp. 813–833, 2020.
2. P. Samal, P. R. Vundavilli, A. Meher, and M. M. Mahapatra, "Recent progress in aluminum metal matrix composites: A review on processing, mechanical and wear properties," *J. Manuf. Process.*, vol. 59, pp. 131–152, 2020.
3. W. Wang et al., "Microstructure and tribological properties of SiC matrix composites infiltrated with an aluminium alloy," *Tribol. Int.*, vol. 120, pp. 369–375, 2018.
4. P. Samal, P. R. Vundavilli, A. Meher, and M. M. Mahapatra, "Influence of TiC on dry sliding wear and mechanical properties of in situ synthesized AA5052 metal matrix composites," *J. Compos. Mater.*, vol. 53, no. 28–30, pp. 4323–4336, 2019.
5. J. Zheng, Q. Li, W. Liu, and G. Shu, "Microstructure evolution of 15 wt% boron carbide/aluminum composites during liquid-stirring process," *J. Compos. Mater.*, vol. 50, no. 27, pp. 3843–3852, 2016.
6. K. Ravikumar, K. Kiran, and V. S. Sreebalaji, "Characterization of mechanical properties of aluminium/tungsten carbide composites," *Measurement*, vol. 102, pp. 142–149, 2017.

7. P. Samal, P. R. Vundavilli, A. Meher, and M. M. Mahapatra, "Fabrication and mechanical properties of titanium carbide reinforced aluminium composites," *Mater. Today Proc.*, vol. 18, no. 7, pp. 2649–2655, 2019.

8. P. Samal, R. K. Mandava, and P. R. Vundavilli, "Dry sliding wear behavior of Al 6082 metal matrix composites reinforced with red mud particles," *SN Appl. Sci.*, vol. 2, no. 2, p. 313, 2020.

9. J. F. Shackelford, Y.-H. Han, S. Kim, and S.-H. Kwon, *Materials Science and Engineering Handbook*, 4th ed. CRC Press, 2015.

10. W. Y. Zhang, Y. H. Du, and P. Zhang, "Vortex-free stir casting of Al-1.5 wt% Si-SiC composite," *J. Alloys Compd.*, vol. 787, pp. 206–215, 2019.

11. K. Amouri, S. Kazemi, A. Momeni, and M. Kazazi, "Microstructure and mechanical properties of Al-nano/micro SiC composites produced by stir casting technique," *Mater. Sci. Eng. A*, vol. 674, pp. 569–578, 2016.

12. T. B. Rao, "An experimental investigation on mechanical and wear properties of Al7075/SiCp composites: effect of SiC content and particle size," *ASME J. Tribol.*, vol. 140, pp. 0316011-8, 2018.

13. A. Rehman, S. Das, and G. Dixit, "Analysis of stir die cast Al-SiC composite brake drums based on coefficient of friction," *Tribol. Int.*, vol. 51, pp. 36–41, 2012.

14. G. B. Veeresh Kumar, C. S. P. Rao, and N. Selvaraj, "Studies on mechanical and dry sliding wear of Al6061-SiC composites," *Compos. Part B Eng.*, vol. 43, no. 3, pp. 1185–1191, 2012.

15. S. Mosleh-Shirazi, F. Akhlaghi, and D. Li, "Effect of SiC content on dry sliding wear, corrosion and corrosive wear of Al/SiC nanocomposites," *Trans. Nonferrous Met. Soc. China*, vol. 26, pp. 1801–1808, 2016.

16. S. Ozden, R. Ekici, and F. Nair, "Investigation of impact behaviour of aluminium based SiC particle reinforced metal-matrix composites," *Compos. Part A Appl. Sci. Manuf.*, vol. 38, pp. 484–494, 2007.

17. K. Bandil et al., "Microstructural, mechanical and corrosion behaviour of Al–Si alloy reinforced with SiC metal matrix composite," *J. Compos. Mater.*, vol. 53, no. 28–30, pp. 4215–4223, 2019.

18. R. K. Singh, A. Telang, and S. Das, "Microstructure, mechanical properties and two-body abrasive wear behaviour of hypereutectic Al—Si—SiC composite," *Trans. Nonferrous Met. Soc. China (English Ed.*, vol. 30, pp. 65–75, 2020.

19. H. Yang, T. Gao, Y. Wu, H. Zhang, J. Nie, and X. Liu, "Microstructure and mechanical properties at both room and high temperature of in-situ TiC reinforced Al–4.5Cu matrix nanocomposite," *J. Alloys Compd.*, vol. 767, pp. 606–616, 2018.

20. K. Ravi Kumar, K. Kiran, and V. S. Sreebalaji, "Micro structural characteristics and mechanical behaviour of aluminium matrix composites reinforced with titanium carbide," *J. Alloys Compd.*, vol. 723, pp. 795–801, 2017.

21. P. Samal and P. R. Vundavilli, "Investigation of impact performance of aluminum metal matrix composites by stir casting," *IOP Conf. Ser. Mater. Sci. Eng.*, vol. 653, p. 012047, 2019.

22. S. Gopalakrishnan and N. Murugan, "Production and wear characterisation of AA 6061 matrix titanium carbide particulate reinforced composite by enhanced stir casting method," *Compos. Part B Eng.*, vol. 43, no. 2, pp. 302–308, 2012.

23. A. Kumar, M. M. Mahapatra, and P. K. Jha, "Fabrication and characterizations of mechanical properties of Al-4.5%Cu/10TiC composite by in-situ method," *J. Miner. Mater. Charact. Eng.*, vol. 11, pp. 1075–1080, 2012.

24. M. S. Song, M. X. Zhang, S. G. Zhang, B. Huang, and J. G. Li, "In situ fabrication of TiC particulates locally reinforced aluminum matrix composites by self-propagating reaction during casting," *Mater. Sci. Eng. A*, vol. 473, pp. 166–171, 2008.

25. K. R. Ramkumar, S. Sivasankaran, F. A. Al-mufadi, S. Siddharth, and R. Raghu, "Investigations on microstructure, mechanical, and tribological behaviour of AA 7075–x wt.% TiC composites for aerospace applications," *Arch. Civ. Mech. Eng.*, vol. 19, no. 2, pp. 428–438, 2019.

26. S. V. Sujith, M. M. Mahapatra, and R. S. Mulik, "An investigation into fabrication and characterization of direct reaction synthesized Al-7079-TiC in situ metal matrix composites," *Arch. Civ. Mech. Eng.*, vol. 19, pp. 63–78, 2019.

27. S. V. Sujith, M. M. Mahapatra, and R. S. Mulik, "Microstructural characterization and experimental investigations into two body abrasive wear behavior of Al-7079/TiC in-situ metal matrix composites," *Proc. Inst. Mech. Eng. Part J J. Eng. Tribol.*, vol. 234, no. 4, pp. 588–607, 2020.

28. S. Arivukkarasan, V. Dhanalakshmi, B. Stalin, and M. Ravichandran, "Mechanical and tribological behaviour of tungsten carbide reinforced aluminum LM4 matrix composites," *Part. Sci. Technol.*, vol. 36, no. 8, pp. 967–973, 2018.

29. D. S. E. J. Dhas, C. Velmurugan, K. L. D. Wins, and K. P. Boopathiraja, "Effect of tungsten carbide, silicon carbide and graphite particulates on the mechanical and microstructural characteristics of AA 5052 hybrid composites," *Ceram. Int.*, vol. 45, pp. 614–621, 2019.

30. U. K. Annigeri and G. B. V. Kumar, "Physical, mechanical, and tribological properties of Al6061-B4C composites," *J. Test. Eval.*, 2019.

31. K. Kalaiselvan, N. Murugan, and S. Parameswaran, "Production and characterization of AA6061–B4C stir cast composite," *Mater. Des.*, vol. 32, pp. 4004–4009, 2011.

32. A. Canakci, "Microstructure and abrasive wear behaviour of B4C particle reinforced 2014 Al matrix composites," *J. Mater. Sci.*, vol. 46, pp. 2805–2813, 2011.

33. B. Manjunatha, H. B. Niranjan, and K. G. Satyanarayana, "Effect of mechanical and thermal loading on boron carbide particles reinforced Al-6061 alloy," *Mater. Sci. Eng. A*, vol. 632, pp. 147–155, 2015.

34. L. Gomez, D. B. Matrix, V. Amigo, and M. D. Salvador, "Analysis of Boron Carbide Aluminum Matrix Composites," *J. Compos. Mater.*, vol. 43, no. 9, pp. 987–995, 2009.

35. M. F. Ibrahim, H. R. Ammar, A. M. Samuel, M. S. Soliman, and F. H. Samuel, "On the impact toughness of Al-15 vol.% B4C metal matrix composites," *Compos. Part B*, vol. 79, pp. 83–94, 2015.

36. F. Toptan, A. Kilicarslan, A. Karaaslan, M. Cigdem, and I. Kerti, "Processing and microstructural characterisation of AA 1070 and AA 6063 matrix B4Cp reinforced composites," *Mater. Des.*, vol. 31, pp. S87–S91, 2010.

37. A. V Pozdniakov, V. S. Zolotorevskiy, R. Y. Barkov, A. Lotfy, and A. I. Bazlov, "Microstructure and material characterization of 6063/B4C and 1545K/B4C composites produced by two stir casting techniques for nuclear applications," *J. Alloys Compd.*, vol. 664, pp. 317–320, 2016.

38. I. Kerti and F. Toptan, "Microstructural variations in cast B4C-reinforced aluminium matrix composites (AMCs)," *Mater. Lett.*, vol. 62, pp. 1215–1218, 2008.

39. K. M. Shorowordi, T. Laoui, A. S. M. A. Haseeb, J. P. Celis, and L. Froyen, "Microstructure and interface characteristics of B4C, SiC and Al2O3 reinforced Al matrix composites : a comparative study," *J. Mater. Process. Technol.*, vol. 142, pp. 738–743, 2003.

40. K. B. Khan, T. R. G. Kutty, and M. K. Surappa, "Hot hardness and indentation creep study on Al-5% Mg alloy matrix-B4C particle reinforced composites," *Mater. Sci. Eng. A*, vol. 427, pp. 76–82, 2006.

2 Study on Characteristic Evaluation and Industrial Applications of Metal and Nano-Metal Matrix Composites

T. Vishnu Vardhan, Balram Yelamasetti,
and D. V. V. Pavan Kumar
Anantha Lakshmi Institute of Technology and Sciences

CONTENTS

2.1 INTRODUCTION

Metal matrix composites (MMCs) have their prosperous in low-end and high-end applications. Due to their role in automobile industry, space technologies, and in other prominent areas, MMCs are termed as future materials and the most promising materials. The two phases of MMCs are matrix and reinforcement, and on proper fabrication at appropriate selected parameters, they offer excellent properties. In general, the most common materials and alloys of Al, Mg, and Cu in the matrix phase of MMCs are preferred due to their lightweight and the ease they offer for MMC fabrication due to their low melting point. Al alloys are highly preferred in many applications because along with above-mentioned properties, they also possess demanding properties like high ductility, good corrosion properties, high specific strength, wear resistance, and lower coefficient of thermal expansion.

The key ingredients used in the fabrication of MMCs are of various types such as fibers/whiskers/particulates. Most of the researchers [1–5] conducted experiments

DOI: 10.1201/9781003345466-2

and concluded that utilization of long fibers in MMCs with continuous reinforcement has offered superior properties than the remaining. Continuous reinforcements exhibit isotropic properties than the other types. However, high production cost and difficult secondary operations in transforming them to definite shapes limit the usage of long fibers in industrial applications. Due to this reason, the researchers carried their work on MMCs with discontinuous fibers for reinforcement. Discontinuous reinforcements include short fibers, or particulates, and they are readily accepted as they are simple and available at less cost. The thermal and wear properties associated with discontinuous MMCs are superior to continuous MMCs [6, 7]. The expected strength of MMCs is influenced by the nature of particulate, whether it is soft or hard, size, volume fraction, and its distribution in the alloy. The selection of stiffer particulates in the reinforcement in soft matrix materials leads to improvement of the properties. For Al matrix, the nitrides, carbides, borides, and oxides are used as stiffer particulate reinforcements to enhance the properties [8]. Casting and powder metallurgy methods are the most common methods for adding reinforcements such as alumina, silicon carbide, etc. [9–12]. In powder metallurgy method, at low temperatures, reinforcement materials in the form of powder and matrix are thoroughly mixed followed by compacting and sintering the same. Very less chances are there for the chemical interaction to take place between matrix and particulates and hence no third product formation takes place in this process. Hence powder metallurgy method is preferred for producing MMCs with high volume particulate ratio [13]. The ball mill is also preferred for production of reinforcements and matrix alloys. Sintering and wet mixing are preferred to obtain uniform distribution of particulates with good mechanical properties, and no secondary operations are required [14–17]. However, the powder metallurgy method has some limitations such as poor wettability and dispersion. This is due to no interaction between metal powders and matrix in liquid phase.

Kevorkijan et al. [18] utilized a nomenclature system by matrix alloy designation, reinforcement, volume %, reinforcement morphology, and generic temper designation to describe aluminum-matrix composites. The morphology indication has been done with 'w' and 'p' for whisker and particulate, respectively. Al6061 with 40% by volume of SiC (p type) processed in a T6 heat treatment (6061/SiC/4Op-T6) has been selected for testing their suitability in producing various automobile components. The gradual application of MMCs is quite interesting, as they offer supporting results and are suitable for high-volume automobile production as per customer needs at no compromise in meeting regulatory standards.

Arsenault et al. [19] conducted experiments to determine the strengthening effect of SiC in Al6061 alloy matrix. The tensile tests followed by SEM investigations have been carried out to determine the morphology of SiC, voids, and aspect ratio (length-to-thickness) of platelets. The experiments revealed that the strength level is attributable to the presence of SiC either in platelets or in fibers, and the results are superior to continuum mechanics theory proposals. The thermal expansion coefficient values of SiC and aluminum and their differences resulted in high density dislocation in matrix where as it was resulted as small sub-grain size in Al6061 alloy.

Aylor et al. [20] determined the pitting behavior of Gr/Al and SiC/Al MMCs. The specimens are processed using powder metallurgy. The pitting attack on SiC/Al has been uniformly distributed and pit penetrations were observed up to lesser depths. In Gr/Al, the presence of graphite has given no electro positive shift and provided a significant resistance to passive film breakdown. Due to this, Gr/Al have performed very poorly in marine applications.

Doel et al. [21] conducted tensile tests on SiC alloy and monolithic materials based MMCs at room temperatures. Different particulate sizes such as 5, 13, and 60 pm were selected to carry out the experiments. Along with this the ageing effects on tensile properties are also determined. The ageing of matrix at peak, under-aged, and over-aged are considered and determined that the 5 pm and 13 pm particle MMCs resulted with tensile strength and 0.2% proof stress than unreinforced. In case of 60 pm particles based MMC the results have shown 0.2% reduction in proof stress and tension strength for peak and under-aged conditions and 0.2% greater proof stress in over-aged conditions. All the three variations have shown lower ductility when compared to unreinforced samples. The internal damage due to particle fracture accumulations may cause for voids formation and leads to ductility reduction.

Dolatkhah et al. [22] used friction stir processing for Al5052 sheets with SiC particles for producing MMCs. The effect of TRS (tool rotational speed), number of passes, traverse speed, and directional shifts on the particle sizes was determined. The microstructure, wear, and microhardness properties were determined with the variation of SiC particles to determine optimum conditions for TRS and traverse speed with the main objective of obtaining powder dispersion in MMC. The change of tool traverse direction and passes variations have decreased SiC particles with enhanced hardness number and wear properties.

Emamy et al. [23] introduced a new technique to reveal the TiB_2 particles formation through molten master alloy mixing. The master alloys are Al-4B and Al-8Ti in Ti:B with 5:2 ratio by weight. The TiB_2 formation in-situ conditions through hot stage microscope, SEM, and XRD analysis was determined. The experiments have revealed the presence of TiB_2 particles via boron atoms in TiAl3 particle interfaces.

Doncel et al. [20, 24] investigated the tensile creep behavior of Al-SiC MMC in the range of 230°C–250°C. The plastic flow in lattice diffusion controlled creep dislocation in Al matrix. A creep relation assessment has been done for metals it added threshold stress. In Al-SiC, the threshold stress for creep is not a thermal activation process. The threshold stress for whisker composites is higher than for particulates.

Shu et al. [25] highlighted the importance of bio-inspired nano-carbon reinforced MMCs fabrication to achieve the bio-inspired composite. The investigation is carried on reinforcement and interface mechanisms including design principles. These bio-inspired nano-carbon reinforced MMCs exhibit excellent electrical, thermal, and mechanical properties. The high strength to toughness ratio makes them special and application prospect through increased strength at low toughness. The fabrication techniques such as electronic adsorption, ball milling, and vacuum filtration can be used as focused tools to produce them. The orderly and uniform dispersion arrangement of nano-carbon in desired degree can be easily obtained through mechanical/electronic/molecular level deposition. The wettability along with chemical activity between matrix and reinforcements will affect the composites' interfacial bonding. This leads to infiltration interface when different phases have high wettability and this may leads to interface reactions.

TABLE 2.1

Tensile Test Results at Different Al-MMCs [26]

Composite	Fabrication Technique	UTS (MPa)	Elongation (%)	0.2% PS (MPa)	Young's Modulus (GPa)	Fracture Toughness (J)
Al-Cu	Squeeze Cast (SC)	261	14	174	70.5	—
Al-Cu+Al_2O_3 ($V_f = 0.2$)	SC	374	2.2	238	95.4	—
Al-Cu-Mg (T6)	Spray-Formed Sheet (SFS)	482	10.2	432	73.8	—
Al-Cu-Mg SiC (T6) (V_f 0.1,10mm particle)	SFS	484		437	93.8	—
Al-Cu-Mg + SiC (T4) ($V_f = 0.17$, 3 μm particle)	Powder-Rolled Plate (PRP)	610	8.0	420	99.3	18
Al-Cu-Mg (T4) (2124)	PRP	525	11.0	360	72.4	—
Al-Cu-Mg (T6), (2126)	PRP	474	8.0	425	73.1	26
Al-Cu-Mg + SiC (T6) ($V_f = 0.17$, 3 mm particle)	PRP	590	4.0	510	99.6	17
Al-Si-Mg + SiC (T6), $V_f = 0.1$, 10 μm particle	SRS	343	3.8	321	91.9	—
Al-Si-Mg (T6), (6061)	SRS	264	12.3	240	69.0	—
Al-Zn-Mg-Cu (T6), (7075)	Spray-Formed Extrusion (SFE)	659	11.3	617	71.1	—
Al-Li-Cu-Mg (T6), (8090)	SFS	505	6.5	420	79.5	38
Al-Li-Cu+SiC (T6), $V_f = 0.17$, 3 μm particle	SFS	550	2.0	510	104.5	—
Al-Zn-Mg-Cu+SiC, (T6) ($V_f = 0.12$, 10 μm particle)	SFE	646	2.6	597	92.2	—

2.2   PROPERTIES OF MMCs

Table 2.1 provides the information about various Al-based MMCs, their fabrication techniques, and their tensile properties.

2.3   NANO-PARTICLE METAL MATRIX COMPOSITES

Nano-scaled reinforcement in composite matrix alloys offers excellent properties at elevated and even at room temperatures. Nano-particle metal matrix composites (NMCs) also offer sound properties like good fatigue, creep and high wear resistance [27]. For high-end applications, NMCs are produced by various fabrication techniques such as spray cast technique, compo-cast technique etc. The primary deviation for NMCs from bulk materials is to offer the functional properties and they have laid heavy impact by offering excellent properties suitable for various industrial applications. The inorganic nano-particles in ceramics/metals are preferred than micro/macro particles, as their dispersion influences material properties such as chemical, mechanical, and thermal [28–30]. The unique nature of NMCs is due to the addition of very small quantity of reinforcement for excellent improvement in desired properties, and this behavior of NMCs has shown their importance in various industrial applications [27–34]. For instance, the selection of nano-phased layered silicates/nano-clay has attracted the researchers' attention due to their significant aspect ratio [35–38].

The following are the key areas of research in the field of NMMCs:

- Controlled and highly precise nano-phase production
- Determination of nano-phase properties like chemical and physical properties at different phases such as pre-mixing, mixing, and post-mixing
- Characterization of nano-filled composites to determine mechanical, wear, thermal, corrosion, and electrical properties
- Mathematical modeling and critical computational models for investigation of response of nano-composites at various loads and conditions

Some reinforcements extensively used in MMCs are shown along with their aspect ratios and diameters in Table 2.2.

TABLE 2.2
Typical Reinforcements Used in Metal Matrix Composites

MMC Type	Diameter(μm) and Aspect Ratio	Examples
Particulate	1–25 & 1–4	B_4C, WC, Al_2O_3, TiC, BN
Whisker	0.1–25 & 10–1,000	$Al_2O_3 + SiO_2$, Al_2O_3, SiC
Continuous	3–150 & >1,000	NbTi, SiC, Al_2O_3, $Al_2O_3 + SiO_2$, C, B, W

2.4   CASE STUDY

The following case describes about fabrication, characterization, and microstructural analysis of LM13-nanoZrO$_2$ MMC. The chemical composition of the Al alloy contains various elements in percentage of weight viz. Fe – 1.0, Zn –0.5, Si – 1.2, Mn – 0.5, Ni – 1.5, Mg – 1.4, and Al – bal. The composite was fabricated through stir casting method where the reinforcement was done with 0%–10% by weight variation. Initially, LM13 (aluminum alloy) melting was done in an electrical resistance temperature control furnace at 735°C. The preheated (350°C) reinforcement particles are added in the alumina crucible furnace through continuous stirring action. The mechanical stirrer attached to electrical furnace was used for controlled stirring action. A pool of homogeneous nano ZrO$_2$ in LM13 was then transferred at 708°C into the mold and allowed for natural curing at atmospheric conditions. The castings were tested with ultrasonic flaw detector casting defects and the proper samples were cut as per ASTM standard to conduct experiments. The specimens used for characterization were prepared as-cast nano-composites of LM13 with nano ZrO$_2$ developed by stir cast method were sliced to 17 mm diameter. The extrusion of as-cast LM13 alloy and LM13 with nano ZrO$_2$ MMCs was done by 25% reduction. The extrusion process is employed with pre-heating up to 500°C for the duration of 30 minutes. This process is employed to enhance the mechanical and wear belongings of the fabricated MMC. Artificial ageing was done on MMC samples by using automatic muffle furnace at 175°C for durations of 2, 4, 6, 8, and 10 hours [39–41].

2.5   MICROSTRUCTURAL STUDY

The optical microscope and SEM studies conducted on LM13 MMCs containing 12 wt.% reinforcement have shown the particulate reinforcement distribution and bonding between matrix and reinforcement. The optical and SEM micrographs of chilled nano metal matrix composites (CNMMC) having 12 wt.% are shown in Figures 2.1 and 2.2.

The metallurgical SEM studies are followed by the evaluation of properties of CNMMC at 12% MMCs through tensile, microhardness, compression, fracture toughness, sliding wear, thermal conductivity, and thermal expansion

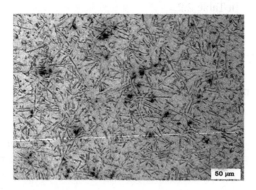

FIGURE 2.1   Optical micrographs of LM13 containing 12% reinforcement [42].

coefficient tests. Overall heat treatment has shown tremendous improvement in all properties at 12% of reinforcement.

The ASTM standards mentioned in Table 2.3 were used in each of the above tests.

FIGURE 2.2   SEM micrograph of LM13 containing 12 % reinforcement [42].

TABLE 2.3
Various ASTM Standards for Testing MMCs

Test Type	ASTM Standard	Description
Microhardness of MMC	ASTM E3	Test load 5 N
Tensile	ASTM E8/95	Specimen actual dimensions 12.5 mm × 62.5 mm/TUE C-400 computerized UTM
Compression	ASTM E9-95	Actual diameter of 13 mm × gauge length of 25 mm /TUE C-400 computerized
Fracture toughness	ASTM E399-1990	Three-point bend test
Sliding wear	ASTM G99-95	Pin-on-disc test rig/room temperature/sample size 8 mm nominal diameter × gauge length of 30 mm/applied load 20, 10, 60, and 80 N at sliding velocities of 1.15, 1.72, 2.3, and 2.88 m/s for a sliding distances of 1037, 2074, 3111, and 4147 m
Thermal conductivity	ASTM E-1461	Micro Flash Laser apparatus/diameter 10 mm and 6 mm height/operating temperature for 100, 200 and 300°C
Thermal expansion coefficient	ASTM E-831	Dilatometer of L75H series apparatus/ test samples of 10 mm diameter × 20 mm height/ operating temperature at 50, 100, 200, and 300°/linear expansion computation through digital indicator

TABLE 2.4

Details of MMC Used for Making LMV Drive Shaft

MMC Type	Property
AlMg1SiCu + 20 vol% Al_2O_3/die casting and extrusion	Stiffness – 95 GPa Fatigue strength – 120 MPa at $n = 50,000,000$ and room temperature Toughness – 21.5 MPa.m1/2 Density – 2.95 g/cc
AlCu4Mg2Zr + 15 vol% SiC particles	Dynamic stability – 100 GPa, yield strength – 413 MPa Fatigue strength (240 MPa for $n = 50,000,000$ at room temperature) Toughness –19.9 MPa.m1/2 Density – 2.8 g/cc

2.6 INDUSTRIAL APPLICATIONS OF MMCs

MMCs have shown their involvement successfully in making automotive engine components such as piston rods, pins, valve trains, engine cover, cylinder head, main bearing, engine block, and others. Heavy iron and aluminum alloys were preferred earlier for making engine components but they have limitations. MMCs are successful in replacing them effectively as they can work successfully at high temperatures. Transverse control arms, brake disks, and housing covers for electronic devices in automotive industry are being produced with MMCs exhibiting high stiffness, high specific strength, small thermal coefficient of expansion, high conductivity, and high thermal resistance [43]. For example, AlMg1SiCu + 20 vol% Al_2O_3 particles used for making light-load vehicle's drive shaft, processed through die casting and extrusion, have shown high stiffness, high dynamic stability, high fatigue strength, and sufficient toughness with low density. Table 2.4 indicates the details of above properties of AlMg1SiCu + 20 vol% Al_2O_3 as a substitution of steel used for manufacturing drive shaft of LMV. Another application of disk of vented passenger car brake fabricated by G-AlSi12Mg + 20 vol% SiC instead of cast iron has been presented. This MMC was produced by sand die casting/gravity die casting. G-AlSi12Mg + 20 vol% SiC possess high wear resistance and low heat conductivity. The longitudinal stringers/bracing beams are produced by using AlCu4Mg2Zr + 15 vol% SiC particles as replacement for PMC for the same application. Die casting, extrusion, and forging processes are employed in making MMC [43–45]. The properties mentioned in Table 2.4 indicate its effective utilization in manufacturing components which require specific properties.

Table 2.5 provide the details of various MMCs, their properties, application areas, and processing techniques.

TABLE 2.5
Applications of Different MMCs

Material System	Required Properties	Applications	Processing Technique
Al-Al$_2$O$_3$, Mg-SiC, Al-SiC, Mg-Al$_2$O$_3$ discontinuous reinforcements	High strength and stiffness, lower thermal expansion coefficient, temperature resistance, thermal conductivity, wear resistance	Automotive industries and heavy goods vehicle, piston rods, frames, piston pins, bracing system, disc brake caliper, Cardan shaft, brake pads	Fusion infiltration, forging, extrusion, die casting, gravity die casting
Al-SiC, Al-C, Al-Al$_2$O$_3$, Al-B, Mg-C, discontinuous and continuous reinforcements	High strength at high temperatures and stiffness, thermal conductivity, lower linear expansion coefficient	Aerospace industry frames, joining elements, reinforcements, aerials	Fusion infiltration, diffusion welding process and joining, extrusion
Al-B, Al-C, Ti-SiC, Al-Al$_2$O$_3$, Al-SiC, Mg-C, Mg-Al$_2$O$_3$, continuous reinforcements and discontinuous reinforcements	High specific strength at elevated temperature and stiffness, temperature resistance, fatigue resistance, impact strength	Military and civil air travel Axel tubes, blades and gear casing, fans and compressor	Fusion infiltration, diffusion welding and soldering, hot pressing, extrusion, squeeze-casting
Pb-Al$_2$O$_3$, PbC	High stiffness, creep resistance	Accumulator plate	Fusion infiltration
Cu-C, Ag-C, Ag-Al$_2$O$_3$, Ag-SnO$_2$, Ag-Ni	Corrosion resistance, high electrical conductivity, burn-up resistance	Electrical contacts	Powder metallurgy, extrusion, pressing, fusion infiltration
Cu-C	High thermal and electrical conductivity, wear resistance	Electrical components, conducting carbon brushes	Fusion infiltration, powder metallurgy
Brass-Teflon	Wear resistance	Bearings	Infiltration
Pb-C	Load carrying capacity		Powder metallurgy
Cu-W	Burn-up resistance	Other applications such as spot welding electrodes	Powder metallurgy, infiltration
Cu-Nb, Cu-Nb3Sn, Cu-YBaCO	Superconducting, mechanical strength, ductility	Super conductor	Extrusion, powder metallurgy, coating techniques

2.7 CONCLUSION

This chapter has dealt with various MMCs and studied about their fabrication, characterization, specific industrial applications, and required properties. The following conclusions are made from this study:

- Thorough inclusion of nano-particles in MMCs has shown improvement in various properties. This has led the focus of researchers to investigate the nano-particle inclusion and determine the need-based MMCs for various industrial applications.
- The case study presented in the chapter will be helpful in identifying appropriate methodology for selection, fabrication and characterization of MMCs. LM13 MMCs containing nano-particles have shown particulate reinforcement distribution and bonding between matrix and reinforcement.
- The applications of AlMg1SiCu MMCs of 20 vol% Al_2O_3/die casting and extrusion and 15 vol% SiC particles were studied.

REFERENCES

1. Hashim J, Looney L, Hashmi MSJ. Particle distribution in cast metal matrix composites - Part I. *J Mater Process Technol* 2002; 123: 251–257.
2. Siddesh Kumar NG, Shiva Shankar GS, Basavarajappa S, et al. Some studies on mechanical and machining characteristics of Al2219/n-B4C/MoS2 nano-hybrid metal matrix composites. *Meas J Int Meas Confed* 2017; 107: 1–11.
3. Shuvho MBA, Chowdhury MA, Kchaou M, et al. Surface characterization and mechanical behavior of aluminum based metal matrix composite reinforced with nano Al2O3, SiC, TiO_2 particles. Chem Data Collect; 28. Epub ahead of print 2020.
4. AbdulqaderAl-maamari AE, AsifIqbal AKM, Nuruzzaman DM. Mechanical and tribological characterization of self-lubricating Mg-SiC-Gr hybrid metal matrix composite (MMC) fabricated via mechanical alloying. *J Sci*; 10 September 2020.
5. Flanagan S, Main J, Lynch P, et al. A mechanical evaluation of an overaged aluminum metal-matrix-composite (2009 Al/SiC/15p MMC), *Procedia Manuf* 2019; 34: 58–64.
6. Pramanik, A., Basak, A. K., Littlefair, G., Dixit, A. R., & Chattopadhyaya, S. Stress in the interfaces of metal matrix composites (MMCs) in thermal and tensile loading. In *Interfaces in Particle and Fibre Reinforced Composites* (pp. 455–471). Woodhead Publishing, 2020, p. 6112.
7. Tweed JH, Manufacture of 2014 aluminium reinforced with SiC particulate by vacuum hot pressing. *Mater Sci Eng A* 1991; 135(30): 73–76.
8. Löfvander JPA, Dary F-C, Ruschewitz U and Levi CG. Evolution of a metastable FCC solid solution during sputter deposition of Ti-Al-B alloys. *Mater Sci Eng A* 1995; 202(1–2): 188–192.
9. Skolianos S. Mechanical behavior of cast SiCp-reinforced Al-4.5%Cu-1.5%Mg alloy. *Mater Sci Eng A* 1996; 210(1–2): 76–82.
10. Kang GC and Seo HY. The influence of fabrication parameters on the deformation behaviour of the preform of metal-matrix composites during the squeeze-casting processes. *J Mater Process Technol* 1996; 61(3): 241–249.
11. Vicens J, Chédru Mand Chermant JL. New Al–AlN composites fabricated by squeeze casting: interfacial phenomena. *Compos A* 2002; 33(10): 1421–1423.

12. Afonso CRM, Kiminami CS, Bolfarini C, et al. Microstructural characterizatin of spray deposited Al-Y-Ni-Co-Zr alloy and Al-Y-Ni-Co-Zr+SiCp metal matrix composite. *Mater Sci Forum* 2002; 403: 95–100.

13. Lin JT, Bhattacharyya D and Lane C. Machinability of a silicon carbide reinforced aluminium metal matrix composite. *Wear* 1995; 181–183(Part 2): 883–888.

14. Hassan HA, Lewandowski JJ. Effects of particulate volume fraction on cyclic stress response and fatigue life of AZ91D magnesium alloy metal matrix composites. *Mat Sci Eng* 2014; 600: 188–194.

15. Cöcenand Ü and Önel K. The production of Al-Si alloy-SiCp composites via compocasting: Some microstructural aspects. *Mater Sci Eng A* 1996; 221(1–2): 187–191.

16. Kaczmar JW, Pietrzak K and Wlosiński W. The production and application of metal matrix composite materials. *J Mater Process Technol* 2000; 106: 58–67.

17. Tjong SC and Ma ZY. Microstructural and mechanical characteristics of in situ metal matrix composites. *Mater Sci Eng* 2000; 29: 49–197.

18. Kevorkijan VM. Metal matrix composites in the automotive industry. *Metalurgija* 1999; 38: 245–249.

19. Arsenault RJ. The strengthening of aluminum alloy 6061 by fiber and platelet silicon carbide. *Mater Sci Eng* 1984; 64: 171–181.

20. Aylor DM, Moran PJ. Effect of reinforcement on the pitting behavior of aluminum-base metal matrix composites. *Proc–Electrochem Soc* 1984; 84–89: 584–595.

21. Doel TJA, Bowen P. Tensile properties of particulate-reinforced metal matrix composites. *Compos Part A Appl Sci Manuf* 1996; 27: 655–665.

22. Dolatkhah A, Golbabaei P, Besharati Givi MK, et al. Investigating effects of process parameters on microstructural and mechanical properties of Al5052/SiC metal matrix composite fabricated via friction stir processing. *Mater Des* 2012; 37: 458–464.

23. Emamy M, Mahta M, Rasizadeh J. Formation of TiB2 particles during dissolution of TiAl3 in Al-TiB2 metal matrix composite using an in situ technique. *Compos Sci Technol* 2006; 66: 1063–1066.

24. González-Doncel G, Sherby OD. High temperature creep behavior of metal matrix AluminumSiC composites. *Acta Metall Mater* 1993; 41: 2797–2805.

25. Shu R, Jiang X, Sun H, et al. Recent researches of the bio-inspired nano-carbon reinforced metal matrix composites. *Compos Part A Appl Sci Manuf* 2020; 131: 105816.

26. Manna, A.; Bains, H.S.; Mahapatra, P.B. Experimental study on fabrication of Al—Al2O3=Grp metal matrix composites. *J Compos Mater* 2011, 45 (19), 2003–2010.

27. Brabazon D, Browne DJ, Carr AJ, Healy JC. *Proceedings of the Fifth International Conference on Semi-Solid Processing of Alloys and Composites*, 2000, p. 21.

28. Witulski T, Winkelmann A, Hirt G. *Proceedings of the Fourth International Conference on Semi-Solid Processing of Alloys and Composites*. University of Sheffield, UK, 1996, p. 242.

29. Wang W, Ajersch F. *Proceedings of the International Symposium on Advances in Production and Fabrication of Light Metals and MMC*. Alta., Edmonton, Canada, 1992, p. 629.

30. Hashim J, Looney L, Hashmi MSJ. Particle distribution in cast metal matrix composites – Part I. *J Mater Process Technol* 2002; 123(2):251–257.

31. Prangnell PB, Barnes SJ, Withers PJ, Roberts SM. The effect of particle distribution on damage formation in particulate reinforced metal matrix composites deformed in compression. *Mater Sci Eng A* 1996; 220(1–2):41–56.

32. Yotte S, Breysse D, Riss J, Ghosh S. Cluster characterisation in a metal matrix composite. *Mater Charact* 2001; 46 (2–3):211–219.

33. Doel TJA, Bowen P. Tensile properties of particulate reinforced metal matrix composites. *Composites A* 1996; 27 A: 655–665.

34. Bindumadhavan PN, Chia TK, Chandrasekaran M, Wah HK, Lam LN, Prabhakar O. Effect of particle-porosity clusters on tribological behavior of cast aluminum alloy A356- SiCp metal matrix composites. *Mater Sci Eng A* 2001; 315 (1–2): 217–26.
35. Hong SJ, Kim HM, Huh D, Suryanarayana C, Chun BS. Effect of clustering on the mechanical properties of SiC particulate reinforced aluminium alloy 2024 metal matrix composites. *Mater Sci Eng* 2003; A347:198–204.
36. Hemanth J. Development and property evaluation of aluminum alloy reinforced with nano-ZrO2 metal matrix composites (NMMCs). *Mater Sci Eng A* 2009; 507: 110–113.
37. Siddesh Kumar NG, Shiva Shankar GS, Basavarajappa S, et al. Some studies on mechanical and machining characteristics of Al2219/n-B4C/MoS2 nano-hybrid metal matrix composites. *Meas J Int Meas Confed.* 2017; 107: 1–11.
38. Hemanth J. Development and property evaluation of aluminum alloy reinforced with nano-ZrO2 metal matrix composites (NMMCs). *Mater Sci Eng A* 2009; 507: 110–113.
39. Daniel S. Sliding wear behaviour of aluminum alloy. *Composites* 2015; 2: 1–7.
40. Daniel SA. Study on the behaviour of Aluminium alloy (LM13) reinforced with NanoZrO2Particulate. *IOSR J Eng* 2014; 4: 58–62.
41. Miracle DB. Metal matrix composites - From science to technological significance. *Compos Sci Technol* 2005; 65: 2526–2540.
42. Hemanth J, Divya MR. Fabrication and corrosion behaviour of aluminium alloy (LM-13) reinforced with nano-ZrO2 particulate chilled nano metal matrix composites (CNMMCs) for aerospace applications. *J Mater Sci Chem Eng* 2018; 6: 136–150.
43. Hashim J, Looney L, Hashmi MSJ. Particle distribution in cast metal matrix composites - Part II. *J Mater Process Technol* 2002; 123: 258–263.
44. Yang Y, Lan J, Li X. Study on bulk aluminum matrix nano-composite fabricated by ultrasonic dispersion of nano-sized SiC particles in molten aluminum alloy. *Mater Sci Eng A* 2004; 380: 378–383.
45. Chakraborty S, Gupta AK, Roy D, et al. Studies on nano-metal dispersed Cu-Cr matrix composite. *Mater Lett* 2019; 257: 126739.

3 Optimization Studies on Al/LaPO$_4$ Composite Using Grey Relational Analysis

K. Balamurugan
VFSTR (Deemed to be University)

T.P. Latchoumi
SRM Institute of Science and Technology

T. Deepthi and M. Ramakrishna
VFSTR (Deemed to be University)

CONTENTS

3.1 INTRODUCTION

Among the available materials, aluminium was proven to have ductility, electrical conductivity, strength to weight ratio, lightweight, high strength, and high load-withstanding capacity. Infiltration of hard particles like ceramic particles in the aluminium materials will provide improved material properties [1, 2]. Metal matrix composites will have improved material properties and find a significant place in a wide choice of manufacturing and structural applications such as marine, aerospace, military, etc. [3–5]. The primary reason for the addition of ceramic particles is to provide superior strength-weight properties and also to increase the toughness property of composites which significantly improves machining performances [6, 7]. The matrix will control the international property of composite materials and it further affects the temperature to strength ratio. The orientation of the reinforcement particle inside the matrix will significantly determine the property of the composite material to a greater extent. Therefore, it was concluded that apart from the selection of suitable reinforcement for the matrix, the particle size and shape will show an influential factor in the determination of the property of the composite material [8].

The various casting processing parameters, temperature holding time, and addition of reinforcement particles in the vortex significantly determine the production process of the composite. In particular, while manufacturing aluminium-based silicon carbide metal matrix composites these factors predominantly determine the property of the composite materials that are fabricated through the conventional stir casting process [9]. The effect of this parameter at the time of fabrication of the composite materials mainly influences the formation of the intermetallic product that is formed due to the uncontrolled chemical reaction that happens between the reinforcement and the base matrix. The selection of proper weight percentage or volume percentage of reinforcement materials that have to be added into the vortex of the molten mixture will improve the wettability, and this process greatly improves the particle distribution in the matrix. Usually, magnesium with 1%–3% in the volume percent was added in the molten mixture to improve the wettability of the test composite materials inside the matrix and this process of production of the composite provides better distribution of particles in the composite materials [10].

Generally, in the aluminium metal matrix composites, some of the ceramic reinforcements such as silicon carbide, tungsten carbide, silicon nitride, aluminium oxide, and aluminium nitride are used. Many researchers had worked with these ceramic materials in the aluminium matrix composite, and shown the improved composite materials properties like thermal properties and wear-resistant properties [11]. Earlier studies reported that preheating of the reinforcement significantly improves the dispersion rate inside the matrix composites and progresses to the uniform dispersions of the reinforcement particles inside the matrix as well [12]. When ceramic particles are used as reinforcement in the metal matrix composite, the preheating temperature will be varied from 650°C to 900°C and this was pursued based on the particle properties. Some materials such as porous graphites were heated under vacuum conditions to avoid oxidation of the material in the open environment [13].

SiC particles are one of the common reinforcement used for the various grades of aluminium matrix composite. The bond that exist between this reinforcement with the base matrix were reported to the excellent and it finds a wide range of applications in the various fields in the engineering [14]. The selection of reinforcement in the matrix considerably determines the material's property. For example, aluminium oxide which possesses similar properties like SiC, when added as reinforcement in the aluminium metal matrix composites, produces an adverse effect on the material's property. Hence, the addition/selection of alumina ceramic particles in the metal matrix composites had less importance than the other ceramic materials. Owing to improve the property of the composite material, this sort of reinforcement decrements the property and has reduced efficiency [15]. From the above statement, it can be stated that the selection of reinforcement significantly implicates the applications of the composite material. The addition of a different reinforcement in aluminium metal is always a challenging task for the researchers, as the selected reinforcement particles have to improve the material's property rather than decreasing.

The various grades of the aluminium alloy were heated to a temperature of 800°C to change the state from solid to liquid. Reinforcement was preheated to a temperature of 200°C to remove the oxidation content and other foreign particles which may greatly affect the properties of the materials or may lead to chemical reactions and progress to get an intermetallic phase material. Generally, the stirring condition will be performed to a spindle speed of 200 rpm for a span of 10–20 minutes using the motorized stirrer to blend the reinforcement in the molten mixture. The reinforcement was slowly added into the vortex of the molten mixture, as due to the centrifugal force the uniform dispersion of particles occurs in the molten mixture. It is also advised to preheat the stirrer to two-thirds of the temperature in the molten metal. The stirring speed has a major effect while on the blending of reinforcement in the molten mixture. High stirring conditions increase the depth of the vortex and can cause the agglomeration of the particle inside the molten material itself. This action generates an irregular distribution of particles and simultaneously affects the end product property [16].

Usually tensile and yield property of the composite samples are likely to increase with the addition of the reinforcement particles in particular elements such as silicon carbide, aluminium oxide, and silicon nitride. Adversely the ductile property of the material gets decreased because of the addition of these hard ceramic particles and progress to have improved brittleness property [17]. While pouring the molten material into the mould, it is always advisable to take the bottom type of pouring method, as it significantly reduces the formation of impurities, and the presence of some foreign elements in the mixture would significantly affect the casting process when performed in an open environment. The removal of the oxide layer from the molten liquid will probably decrease the chance of getting a defective product and improve the mechanical property of the composite samples [18].

Setting temperature above the melting point of the source material in the matrix has played a vital role in the reduction of the formation of any intermetallic phase elements due to the addition of reinforcement. It is proposed to maintain 30°C–50°C above the base material melting point. Higher temperature may also lead to the evaporation of the material and instant reaction with the elements present in the environment that results in the formation of the slag. Low heating of the mould also plays a significant role in getting the particle dispersion in the matrix. Low temperature heated mould will absorb the excess heat from the molten liquid that was introduced into the mould and cause sudden cooling of materials. This action mainly results in the formation of grain and the grain boundaries for the fabricated composite.

Introducing low-density material to the aluminium matrix composites in particular while fabricating the composite through the stir casting process was a challenging task for the manufacturers and the researchers. Segregation of the composite will be visible due to the gravity attribute during the particulate formation. It is always advisable to add a high-density compound followed by a low-density compound as reinforcement in the aluminium matrix composite [19]. By following methodologies i.e., the addition of high-density reinforcement followed by the low-density reinforcement, the researchers will be able to acquire uniform distribution of particles and overcome the challenges under the traditional stir casting process [20].

LaPO$_4$ had proven that the particles will exhibit an excellent high thermal withstanding capacity, corrosion resistance, chemically inert, and no phase transformation or formation of any intermetallic phase elements even above 1400°C. LaPO$_4$ resists the thermal shock that occurs during the machining and operating condition. This behaviour makes it suitable for candidate materials for various thermal barrier materials, refractory materials, and corrosion resistance material [21]. LaPO$_4$ is a non-reactive compound with ZrO$_2$ up to 1600°C [22]. The addition of LaPO$_4$ forms as an interfacial layer between the grains and while on machining, the ceramic particle progress to the deflection of the material by creating or propagation of cracks [23]. This action enhances the machinability characteristics of the composite materials. No such ceramic reinforcement provides such material characteristics. The above statement was validated by an author on studying the machinable property of Al$_2$O$_3$-LaPO$_4$ ceramic composites. It was reported that increase in the addition of LaPO$_4$ in the matrix will promote the internal microcracks and by using the conventional machining process the ceramics composite can be machined was only because of the presence of LaPO$_4$. Besides the addition of an excess quantity of LaPO$_4$, i.e., more than 40% of the weight ratio in the matrix will reduce the feasibility of the fabricated composite [24].

LaPO$_4$ deposition on the tantalum substrate proves to admirably resist corrosion, as it forms a layer against liquid uranium and improves the life of the samples even at 1573 K [25]. Superior densification will be enhanced when LaPO$_4$ is added to phosphor materials [26]. Excellent phase stability is obtained on the addition of LaPO$_4$ [27]. LaPO$_4$ with zirconia was proven to have reduced hardness observations when compared to the other ceramic particles and it was stated that LaPO$_4$ was soft ceramic comparatively [28, 29].

Particle-reinforced aluminium matrix composite is fabricated with different techniques like powder metallurgy, spray process, squeeze casting, stir casting, and friction stir welding and also with various commercial processing techniques [30]. From the earlier studies, it is confirmed that the foundry-based casting manufacturing techniques were proven to be more viable for aluminium matrix composites and provide a very promising behaviour for the manufacturing firm. The components can be manufactured with less design complexity, less defects and cost-effectively. Sufficient wettability of foreign particles at the liquid state and homogeneous dispersal of these particles are general problems that occur during casting techniques [31]. Some of the unsatisfactory structural defects like agglomeration oxide inclusion white regions interfacial reaction between the components in the mixture occur in the casting process [32]. To overcome these defects, several novel technologies and modifications were adopted for the existing traditional casting process. For instance, the ultrasonication technique, which can be used as an accessory in the stir casting process, produces some small vibration in the form of kilohertz. This vibration produced by the ultrsonificator in the casting process enhances to get the uniform distribution of particles in the matrix [15]. The recent ultrasonication-assisted stir casting process is identified as a suitable fabricating technique because the samples fabricated using this accessory will provide the uniform circulation of reinforcement inside the mixture and in microstructure [33]. As the application of the ultrasonicator, small vibration was introduced into the mixture at the time of casting that enhances the uniform distribution of particles inside the mixture. This action reduces the agglomeration of the particles besides being helpful for uniform orientation of the reinforcement particle in the aluminium matrix composites [34].

Besides the input parameters, the material properties significantly ensure in determining the machinability characteristics of the end material. The abrasive water jet machine (AWJM) as compared to other machining processes has fewer disadvantages such as noisy operation and creating an untidy machining environment. Moreover, the improper selection of the AWJM condition will lead to having reduced machining output with the formation of large kerf taper, which is unavoidable and causes severe defects in AWJM [35, 36], although AWJM is proven as a suitable cutting technology for the through cut operations with no heat-affected zone and free from thermal cracks. Al/LaPO$_4$ composite properties could be used for high-temperature and high-melting applications [37].

Earlier studies used to deal with the decision problems which directly require some huge data such as linear programming and regression analysis. For example, in the real-time problems inefficient samples, instability, and irregularity of data may lead to numerous assumptions and considerably increase the mathematical complexity [38, 39]. Grey relational analysis (GRA) was a proven and absolute approach for any kind of decision-making problem irrespective of the variations in the data. GRA was used to contribute to solving objective problems and complex problems [40–42]. It was intended to be used for the small set of the database that was partially known and aimed to provide the complete information with the available data which makes the GRA as a suitable technique for machining conditions, as it deals with the limited numbers of datasets [43].

In the present work, the Al/15% LaPO$_4$ weight fraction of the metal matrix composite was prepared by the ultrasonic-aided stir casting process. The cast sample is immediately compressed to 2 tons of uniaxial load per square inch. The machinability characteristics of the samples are evaluated using AWJM for the different operating conditions. To find optimal operating condition to the varied input parameters such as jet pressure, standoff distance, and cutting speed for the three output responses such as kerf angle (KA), material removal rate (MRR), and Ra through GRA. The effect of the use of two different abrasives namely silicon carbide and garnet over the kerf surface is analysed through microscopy examination and reported.

3.2 MATERIALS AND METHODS

3.2.1 SAMPLE PREPARATION

To prepare LaPO$_4$ powder, the calculated quantity of lanthanum nitrate was mixed with orthophosphoric acid in an ultrasonic bath. At a pH level of 2, the solution was allowed to settle down. The white precipitate deposited was air-dried, packed together, and heated to obtain $LaPO_4$. To identify the suitable weight percentage of $LaPO_4$ in the aluminium matrix composite, the primary studies were conducted with the varied weight percentage of $LaPO_4$. The percentages varied are with the multiples of 5 (such as 0%, 5%, 10%, 15%, and 20%). The basic mechanical testing such as harness, flexural, and tensile examinations was conducted on the fabricated aluminium-based metal matrix composites. From the observation, it was concluded that $LaPO_4$ with a 15% weight ratio was identified to produce better mechanical characteristics and proposed as a suitable mixture ratio in the aluminium matrix composite. Lower percentage of $LaPO_4$ in the aluminium provides a similar observation as compared with the nascent sample, i.e., 0% of $LaPO_4$. With an increase in the percentage of $LaPO_4$ reinforcement in the aluminium matrix composites, the dispersion of these reinforcement particles in the mixture was recorded to be so complicated. Irrespective of ultrasonication where a frequency of 2 kHz is given to the mixture, agglomeration becomes unavoidable above 15% of lanthanum phosphate as reinforcement in the aluminium matrix composites. Due to the agglomeration, the observation was found to be very low compared to the 15% of reinforcement of lanthanum phosphate in weight percentage.

On microstructural examinations, the fracture surface reveals the presence of voids in 20% $LaPO_4$ and this level of increase in porosity observed in the aluminium composite samples was mainly due to the irregular placement of reinforcement in the matrix. Irrespective of hardness, flexural, and the other basic mechanical observation, the fabricated composite showed the least observations on 20% of reinforcement. Therefore it was concluded that 15% of $LaPO_4$ in the aluminium metal matrix composites was proposed as suitable weight percentage reinforcement in aluminium matrix.

The calculated quality of the Al6061 rod that was procured from Coimbatore Metal Mart, Coimbatore, India was melted in an electrical furnace at about 780°C. After a preliminary investigation, the initially prepared 15% $LaPO_4$ nanopowder with a 15% weight ratio was slowly added into the vortex of the mixture. The composition was stirred for 10 minutes in the molten state using the two flute motorized stirrer that operated at 200 rpm. A die of 120 mm diameter and 10 mm height was preheated to a temperature of 250°C. The molten mixture was allowed to fill the preheated die. The cast sample was then compressed to a uniaxial compression load of 2 tons/sq. inch. The sample was then allowed to cool at room temperature.

3.2.2 EXPERIMENTAL PROCESS

AWJM has various parameters; among them jet pressure, standoff distance, and cutting speed are found to be the dominating parameters at the time of machining any materials [44–47]. The electrical and thermal conductivity of all ceramic materials was proven to be low. As the electron movement between the particles is resisted, it provides no electrical conduction and it greatly resists the thermal transfer between the particles. This action usually reduced the electron and thermal conductivity for any conducting materials [48]. This property of the $LaPO_4$ particles will limit the machine applications. Some electrical and thermal cutting process like

TABLE 3.1

The AWJM Machining Parameters with Levels

S. No.	Factors	1	2	3	Units
			Levels		
1	Jet pressure (JP)	220	240	260	bar
2	Standoff distance (SOD)	1	2	3	mm
3	Cutting speed (CS)	20	30	40	mm/s

electric discharge machining, plasma cutting, laser cutting, etc., was identified to be an inappropriate machining process for the fabricated composite. In the present study, three parameters (JP, SOD, and CS) are identified as independent parameters to measure three dependent parameters. The amount of material removed at each cut-off distance of 15 mm was measured and recorded as the MRR; kerf taper is unavoidable in the abrasive water jet machining process hence KA is also taken as one of the dependent parameters. Each machining operation is measured based on the surface finish of the process. In the present study, the surface profile parameter was taken as one of the measurable parameters. The considered machining parameter and its machining conditions are shown in Table 3.1. The commercially available silicon carbide and garnet that has a mesh size of 80 are used as abrasives.

The quantity of the material removed from the composite during the different machining conditions was measured using Equation 3.1, and the KA was measured by using Equation 3.2:

$$MRR = \frac{(\text{Initial weight} - \text{final weight})}{\text{sample density} \times \text{duration}} \tag{3.1}$$

$$KA = \frac{(\text{Top kerf width} - \text{bottom kerf width})}{2 \times \text{sample height}} \tag{3.2}$$

Mitutoyo made surface roughness tester with model SJ-411 was used to measure the Ra. The surface tester with a measurement range of 350 μm and at a measurement speed of 0.25 mm/s was allowed to move to a length of 5 mm on the kerf surface. A surface profile projector was used to take the KA observation. Shimadzu makes high-precision weighing balance equipment (AUX 220 model) which was used to measure the quantity of materials eroded during the AWJM process.

3.3   RESULTS AND DISCUSSIONS

3.3.1   GREY RELATIONAL ANALYSIS

The main objective of GRA is to prefer linear normalization which undergoes a multi-input type of optimization process to analyse the problem for an efficient solution. Based on the output performance characteristics, the optimistic machining parameters can be determined. The objective is to get the superior Ra with an acceptable level of MRR and KA. Smaller the better is chosen for KA, and Ra and larger the better are chosen for MRR. The S/N ratio represents the two changes of desirable and undesirable values, which depend upon the machining characteristics like surface roughness and wear ratio.

The grey relation technique is the process, relating to the observations that have 'n' numbers of experiment and it is allowed to converge between 0 and 1 for normalizing. This action is created in this technique is to avoid the unit variations and to reduce the variations in the experimental observation.

The S/N ratio can be calculated by using Equations 3.3 and 3.4:

$$\text{Smaller the better}: \frac{S}{N_{SB}} = -10 \log_{10} \left[\frac{1}{n} \left(\sum_{i=1}^{n} y_{ij}^2 \right) \right] [[Tab]][[Tab]] \quad (3.3)$$

$$\text{Larger the better}: \frac{S}{N_{LB}} = -10 \log_{10} \left[\frac{1}{n} \sum_{i=1}^{n} \frac{1}{y_{ij}^2} \right] [[Tab]][[Tab]][[Tab]] \quad (3.4)$$

where y_{ij} is the ith examination on the jth trial and n is the total number of a trial conducted.

$$\text{The larger the best } x_{ij} = \frac{\left[X_{ij} - \min_j y_{ij} \right]}{\left[\max_j y_{iy} - \min_j y_{ij} \right]} [[Tab]][[Tab]][[Tab]] \quad (3.5)$$

$$\text{The smaller the best } x_{ij} = \frac{\left[\max_j y_{ij} - X_{ij} \right]}{\left[\max_j y_{iy} - \min_j y_{ij} \right]} [[Tab]][[Tab]][[Tab]] \quad (3.6)$$

where y_{ij} is ith examination in jth trial.

The interpretation of the grey relation coefficient relation gives the relevance between the ideal best experimental observation values and the normalized experimental observation values.

$$\delta_{ij} = \frac{\min_i \min_j \left| X_i^0 - X_{ij} \right| + \zeta \max_i \max_j \left| X_i^0 - X_{ij} \right|}{\left| X_i^0 - X_{ij} \right| + \zeta \max_i \max_j \left| X_i^0 - X_{ij} \right|} [[Tab]][[Tab]][[Tab]] \quad (3.7)$$

where $x_i^0 =$ ideal normalized observations for the jth trial,
$\zeta =$ coefficient.

The grey relational grade (GRG) will provide the average grade relationship characteristics for each of the corresponding performance characteristics.

$$y_j = \frac{1}{m} \sum \zeta_{ij} \quad (3.8)$$

where $i = 1$ to m
$\gamma_j =$ GRG of jth trial in mth number of performance characteristics.

The grey relation coefficient and GRA is calculated by using Equations 3.7 and 3.8, respectively. The authors had followed the procedure adopted by Noorul Haq et al. [49] and Balamurugan et al. [50–52].

The experimental observations at different work conditions in AWJM with GRG are shown in Table 3.2. Smaller the better characteristics in GRG provide the best acceptable machining conditions for Ra. From the observation, it is found that the first experimental condition on the orthogonal array of JP = 220 bar, SOD = 1 mm, and CS = 20 mm/min gives a better output response characteristic for both the garnet and silicon carbide. Tables 3.3 and 3.4 display the response observations of garnet and SiC, respectively.

TABLE 3.2
Experimental Data

| S. No. | JP (bar) | SOD (mm) | CS (mm/ min) | Garnet | | | | Silicon Carbide | | | |
				MRR (g/s)	KA (°)	Ra (μm)	GRG	MRR (g/s)	KA (°)	Ra (μm)	GRG
1	220	1	20	0.04017	0.141	2.035	0.8120	0.04059	0.296	1.2505	0.7586
2	220	2	30	0.03744	0.189	3.197	0.5334	0.04068	0.413	1.3485	0.5855
3	220	3	40	0.03989	0.256	4.755	0.4089	0.03965	0.529	1.7345	0.4277
4	220	1	30	0.03635	0.153	2.445	0.6791	0.03659	0.366	1.3875	0.5982
5	220	2	40	0.03496	0.207	3.616	0.4690	0.03751	0.462	1.6015	0.4754
6	220	3	20	0.04354	0.202	3.797	0.5324	0.04841	0.389	1.4035	0.5988
7	220	1	40	0.03461	0.193	2.853	0.5388	0.03022	0.455	1.5455	0.4793
8	220	2	20	0.04054	0.173	2.984	0.6013	0.04328	0.349	1.1985	0.7326
9	220	3	30	0.04206	0.227	4.187	0.4709	0.04256	0.453	1.5675	0.4974
10	240	1	20	0.04112	0.182	2.618	0.6225	0.05182	0.383	1.2905	0.6629
11	240	2	30	0.03953	0.213	4.016	0.4706	0.05361	0.5	1.5905	0.4998
12	240	3	40	0.04204	0.281	5.193	0.4007	0.05589	0.608	1.8565	0.4301
13	240	1	30	0.03862	0.204	3.006	0.5332	0.04482	0.455	1.5535	0.5047
14	240	2	40	0.03791	0.246	4.505	0.4100	0.05087	0.545	1.7415	0.4479
15	240	3	20	0.04762	0.225	4.335	0.5526	0.06691	0.506	1.5745	0.5587
16	240	1	40	0.03643	0.227	3.469	0.4627	0.04267	0.502	1.7105	0.4479
17	240	2	20	0.04227	0.193	3.583	0.5418	0.05596	0.384	1.3905	0.6305
18	240	3	30	0.04428	0.239	4.835	0.4647	0.06084	0.559	1.7155	0.4828
19	260	1	20	0.04504	0.2	3.512	0.5701	0.06565	0.439	1.4825	0.6063
20	260	2	30	0.04526	0.236	4.483	0.4922	0.06785	0.553	1.7905	0.5102
21	260	3	40	0.04647	0.299	5.513	0.4455	0.07122	0.702	2.0255	0.4713
22	260	1	30	0.04164	0.219	3.83	0.4900	0.06421	0.497	1.6315	0.5358
23	260	2	40	0.04251	0.28	5.253	0.4047	0.06429	0.645	1.9405	0.4477
24	260	3	20	0.05035	0.255	4.897	0.6082	0.07957	0.607	1.7045	0.6021
25	260	1	40	0.04013	0.267	4.351	0.4164	0.05722	0.583	1.7515	0.4556
26	260	2	20	0.04779	0.223	3.972	0.5728	0.07558	0.484	1.6575	0.6180
27	260	3	30	0.04854	0.277	5.159	0.5127	0.07644	0.689	1.9065	0.5322

The low levels of input parameters are found to produce the optimum machining parameter. The first level in three input parameters suits to fit the requirements. The optimum machining condition while using garnet as abrasive is A1B1C1. Correspondingly, for silicon carbide is D1E1F1.

The variations in stages of individual input parameters over the dependent responses on two different abrasives are shown in Figure 3.1. Figure 3.1a and b shows the relation between the GRG and the levels of garnet and SiC, respectively. It is observed that a steady steep fall in CS on using garnet and SiC as abrasives reveals the greater impact of CS in affecting the output responses than other input parameters.

TABLE 3.3
Response Table of the Functional GRG on Garnet

Symbols	Machining Parameters	Response Table			
		Level 1	Level 2	Level 3	Max-Min
A	JP (bar)	0.5606	0.4954	0.5014	0.0652
B	SOD (mm)	0.5694	0.4995	0.4885	0.0809
C	CS (mm/min)	0.6015	0.5163	0.4396	0.1619
Error		0.5272	0.5263	0.5119	0.5192

TABLE 3.4
Response Table of the Functional Grey Relation Grade on Silicon Carbide

Symbols	Machining Parameters	Response Table			
		Level 1	Level 2	Level 3	Max-Min
D	JP (bar)	0.5726	0.5184	0.5310	0.0542
E	SOD (mm)	0.5610	0.5497	0.5112	0.0498
F	CS (mm/min)	0.6409	0.5274	0.4537	0.1873
Error		0.5587	0.5457	0.5264	0.5499

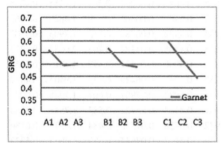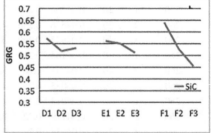

FIGURE 3.1   GRG vs levels.

3.3.1.1 Analysis of Variance

ANOVA for garnet shown in Table 3.5 reveals the influence of CS on the output responses with a contributing percentage of 66.67 followed by SOD with 19.58% and JP with 13.21%. The reduction of the cutting time due to high levels of CS will significantly affect the output responses. It produces waviness and large peaks on the rough cut region. The increase of SOD will increase jet divergence and width of the water beam which will lead to the partial loss of kinetic energy gained by the abrasive particles. At high jet pressure, the collision of abrasives with the backscattered particles partially reduces the tendency and the flow direction of abrasives which has been found to have significantly affected the output performance.

ANOVA for SiC is shown in Table 3.6. The substantial effect of CS in the output responses on using silicon carbide as an abrasive in the machining of the composite is reported with a contribution of 84.41%. Least contributions of JP and SOD are reported with 6.4% and 7.6%, respectively. A considerable increase in CS affects the water flow that influences on the abrasive's flow rate. Lesser MRR with a larger KA and reduced Ra can be obtained with a variation in CS levels; based on the requirements of the end product, the manufacture can select the abrasive irrespective of running cost.

Figure 3.2 provides the percentage of contribution of each input parameter while using garnet and SiC, as abrasives. In AWJM, the determination of the output

TABLE 3.5
Results on ANOVA for Garnet as Abrasive

Parameters	DOF	Sum of Square	Mean Square	Contribution (%)	F-Value	
JP (bar)	2	0.0078	0.0039	13.21	25.1729	Significant
SOD (mm)	2	0.0116	0.0058	19.58	37.2992	Significant
CS (mm/min)	2	0.0393	0.0197	66.67	127.0137	Significant
Error	2	0.0003	0.000155	0.52		
Total	8	0.0590		100.00		

TABLE 3.6
Results on ANOVA for Silicon Carbide as Abrasive

Parameters	DOF	Sum of Square	Mean Square	Contribution Q (%)	F-Value	
JP (bar)	2	0.0048	0.0024	7.63	5.1457	Significant
SOD (mm)	2	0.0041	0.0020	6.46	4.3554	Significant
CS (mm/min)	2	0.0534	0.0267	84.41	56.8799	Significant
Error	2	0.0009	0.000469	1.4841		
Total	8	0.0633		100.00		

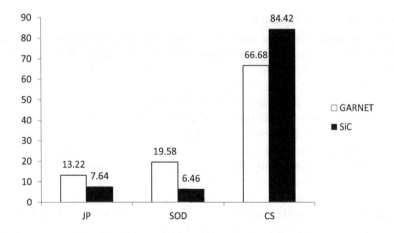

FIGURE 3.2   Contributions of machining parameters.

performance on using garnet and silicon carbide as abrasives greatly depends on CS than jet pressure and standoff distance. The hardness of SiC is found to be superior to garnet. The high accelerated garnet particle, when it impinges on the surface of the composite, loses the energy by removal of material and diffracted to form new small grains. The newly formed grains collide with freshly entered abrasives in the water beam. This makes the abrasives to move in a random path and produces the least output characteristic. A significant rise in JP and SOD is obtained for garnet because the low hard garnet abrasive increases the thickness of the water jet that leads to having a large KA and MRR that consequently affects the surface quality of the composite by creating a microwear track on the entire surface. From the results, the influence of the considered three input parameters have shown a significant effect in affecting the output responses and further, it can be validated through F-test.

3.3.2   Influence of Process Parameters

Using GRA, the output parameters of AWJM are optimized for varied input parameters. On each successive working condition, the influence of one parameter over the other is witnessed in the output performance characteristics. By concluding that on each level of machining parameters, the three varied input machining conditions on two different abrasives have a significant influence in MRR, KA, and Ra. Nearly 86% of the contribution is obtained in CS alone on silicon carbide, as abrasives at different working environments. From Figure 3.3, the SiC, owing to hardness and fracture resistance at high pressure and impact, plunges through the composite and utilizes the acceleration energy in full path in machining the composite. The significant changes in jet pressure and standoff distance determine the acceleration energy of the SiC abrasives. It tends to have a superior observation in MRR and KA. In CS, a significant change in levels irrespective of JP and SOD leads to produce grain wear track and waviness on the rough cut surface.

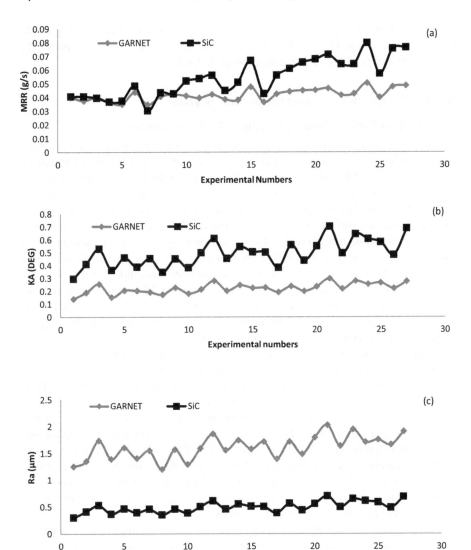

FIGURE 3.3 Image showing the performance deviations of each output response with its input parameters: (a) MRR, (b) KA, and (c) Ra.

It is alleged that a SiC particle passes over the entire thickness of composite by avoiding the cluster formation and collision with backscatter abrasives. It tends to have a least Ra. The low hardness particle shutters itself owing to the applied load and the width of the water beam is found to be less than SiC which results to have low MRR and KA. Multiple hits of backscattered garnet particles on the machined surface affect Ra. The confirmation test is done to verify the improvement in the output characteristics. Final verification is carried out for optimistic working conditions of A1B1C1 and D1E1F1 for the garnet and silicon carbide, respectively. The results obtained are found to be at an acceptable level.

3.3.3 Microscopy Examination on Kerf Surface

3.3.3.1 Kerf Surface Created by Silicon Carbide Abrasive

The hardness property of SiC is found to be higher than the garnet. The high-pressure water beam accelerates this hard SiC abrasive and it is allowed to impinge on the surface of the fabricated composite. Based on the thickness of the sample, the backscattered silicon carbide abrasive will pass over the surface intend to produce some kerf width. The top portion of the surface profile parameter of the silicon carbide machined surface was shown in Figure 3.4a. The SiC abrasive machined surface shows some minor cracks. The simultaneously cooling and heating on the same region may develop thermal stress that cause for small cracks formation on the machined surface. From Figure 3.4b, the uniform distribution of the peak and the valley regions were confirmed. The hardness of the SiC impression is retained even after the primary impingement on the composite surface. The backscattered SiC abrasives create some plastic deformation surface and provide an acceptable range of surface finish with uniform peak and the valley region.

Figure 3.4c is taken in the middle portion of the cut surface. Small worn-out piece stuck over the surface is visible because of the lack of machining time as it reduces the erosion rate as well as the acceleration energy of the abrasive particles. The backscattered electron with the high energy slides over the machined surface and creates a uniform peak and valley region with the acceptable range of surface finish, whereas at the centre part the ratio of the backscattered abrasives compared to the top surface is considerably reduced. The continuous hit of these hard SiC particles on the fabricated composite surface creates crater regions with the removal of an excessive amount of material. Figure 3.4d signifies the spectrum of the centre portion of the cut surface. Random moment of the hard SiC abrasives over the centre portion of the cut surface produces a valley region than the peak.

Figure 3.4e is taken from the bottom surface of the cut region. Excess amounts of wear tracks are found on the machined surface. The linear moment of the hard SiC abrasives creates some curvy direction motion of abrasives and this could be verified from the wear tracks. With the increase in depth of the samples, the newly entered SiC abrasive with high acceleration energy is very limited. It mainly depends upon the cutting speed of the water jet machine. The poor surface finish was recorded at this cutting zone. The variations between the peak and the valley regions are shown in Figure 3.4f. Movement of abrasives over this composite surface and the least amount of scattered particles at this region tend to get a very low surface profile parameter.

3.3.3.2 Kerf Surface Created by Garnet Abrasive

Figure 3.5a shows the top kerf surface of the machined composite. A large number of microcracks are visible in overlapped regions. When garnet strikes the surface

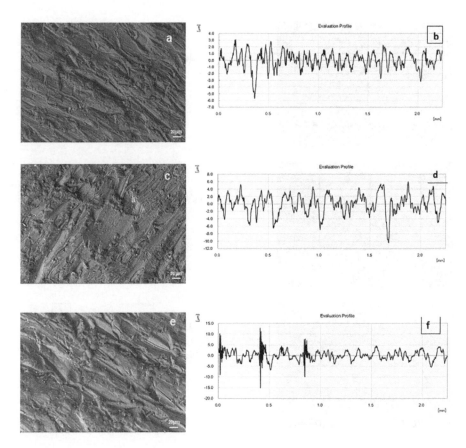

FIGURE 3.4 Al/LaPO$_4$ cut surface using silicon carbide as abrasives. (a) Top cut surface, (b) surface profile of top cut surface, (c) centre surface, (d) surface profile of the centre surface, (e) bottom surface, and (f) surface profile of the bottom surface.

because of the low hardness, kerf surface can be replaced as machined surface or can be retained as such. This is verified by spectrum, and it is shown in Figure 3.5b.

Figure 3.5c shows the middle region of the kerf surface. The repetitive loads on the composite region give a forged effect on the surface with the formation of larger crater regions. The impinged particles, as they pass over the hard composite, tend to tail stack the impression given by the previously passed abrasives and tend to move along the same path. When full energy is gained by the abrasive which hits this surface, the forge effect is produced. Figure 3.5d gives the spectrum of the region which reveals that the formation of Rv is high.

Figure 3.5e shows the bottom cut region of the kerf surface. The large crater with the scar track and the surface with less striation are formed because of the excess quantity of particles removed/eroded during AWJM on the composite by crushing and squeezing actions. This action weakens the grain boundary and causes grain deformation followed by the erosion of particles. Figure 3.5f shows the spectrum

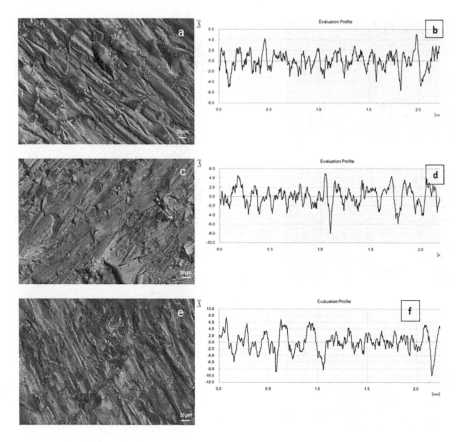

FIGURE 3.5 Al/LaPO$_4$ cut surface using garnet as abrasives. (a) Top cut surface, (b) surface profile of top cut surface, (c) centre surface, (d) surface profile of the centre surface, (e) bottom surface, and (f) surface profile of the bottom surface.

of the bottom kerf surface. When the abrasive is passed over the surface, it slices through the composite leaving behind its impression and leads to getting high Rv.

3.4 CONCLUSION

In this chapter, the aluminium-based composite is fabricated with 15% in a weight ratio of LaPO$_4$ through an ultrasonic-assisted stir casting process. The fabricated Al/LaPO$_4$ composite was machined using an AWJM with the varied machining parameters. The influence of each machining parameter and garnet and SiC abrasive on the curved surface is examined through scanning electron microscopy and discussed, and the observations are as follows:

- SiC and garnet abrasives exhibit optimized conditions at low levels of each input parameter.

- The predominant influence of CS is observed with a contribution of nearly 84.41% and 66.67% for SiC and garnet, respectively.
- The contribution of JP and SOD in SiC is significantly lower than that of garnet. A superior MRR and KA with least Ra are observed for SiC, whereas high Ra with decreased MRR and KA is observed for garnet.
- SiC forms a superior surface finish than garnet because of the hardness property and the high energy backscatter abrasives. SiC shows a contribution of 7.63% and 6.46% for JP and SOD, and garnet 13.21% and 19.58% for JP and SOD, respectively.
- Internal stress and the continuous impingements developed by the hard SiC abrasives lead to grain boundary deformation through the erosion process.
- The kerf surface of SiC and garnet shows the plastic deformation surfaces. SiC abrasives create excess microcracks that leads to the bond failure of the composite whereby crushing and squeezing of the composite materials are witnessed for the garnet abrasives.

REFERENCES

1. Tiryakioglu M, Campbell J. Guidelines for designing metal casting research: Application to aluminium alloy castings. *International Journal of Cast Metals Research*. 2007; 20: 25–29

2. Clyne TW, Withers PJ. *An introduction to Metal Matrix Composites*. Cambridge University Press, Cambridge; 1995. p. 293

3. Sajjadi SA, Ezatpour HR, Beygi H. Microstructure and mechanical properties of Al–Al2O3 micro and nanocomposites fabricated by stir casting. *Materials Science and Engineering A*. 2011; 528: 8765–8771.

4. Miller WS, Zhuang L, Bottema J, Wittebrood AJ, Smet PD, Haszler A, et al. Recent development in aluminium alloys for the automotive industry. *Materials Science and Engineering A*. 2000; 280: 37–49.

5. Brown KR, Venie MS, Woods RA. The increasing use of aluminium in automotive applications. *Journal of the Minerals, Metals and Materials Society*. 1995;47:20–23.

6. Rosso M. Ceramic and metal matrix composites: Routes and properties. *Journal of Materials Processing Technology*. 2006;175:364–375.

7. Xiu ZY, Chen GQ, Wang XF, Wu GH, Liu YM, Yang WS. Microstructure and performance of Al-Si alloy with high Si content by high temperature diffusion treatment. *Transactions of Nonferrous Metals Society of China*. 2010;20:2134–2138

8. Kuhn H, Medlin D. *Mechanical Testing and Evaluation, American Society of Metals ASM Handbook ASM International*. Handbook Committee: USA. 2000;8.

9. Lloyd DJ. The solidification microstructure of particulate reinforced aluminium/SiC composites. *Composite Science and Technology* 1989; 35(2): 159–179.

10. Rohatgi PK, Ray S, Sthena RA and Narendranath CS. Interface in cast metal matrix composites. *Materials Science and Engineering* 1993; 162(1–2): 163–174. doi:10.1016/0921-5093(90)90041-Z.

11. Rohatgi PK, Yarandi FM, Liu Y In: Fishman SG, Dhingra AK (Eds.). *Proceedings of International Symposium on Advances in Cast Reinforced Metal Composites*. ASM International Publication, Materials Park, OH. 1988;249.

12. Etter T, Schulz P, Weber M, Metzc J, Wimmler M, Loffler JF, Uggowitzer PJ. Aluminium carbide formation in interpenetrating graphite/aluminium composites. *Materials Science Engine*, 2007; A 448: 1–6.

13. Soy U, Demir A, Caliskan F. Effect of bentonite addition on fabrication of reticulated porous SiC ceramics for liquid metal infiltration. *Ceramics International*, 2011; 37: 15–19.

14. Surappa MK. Aluminium matrix composites: Challenges and opportunities. *Sadhana* 2003; 28(Parts 1 & 2): 319–334.

15. Hashim J, Looney L and Hashmi MSJ. Metal matrix composites: Production by the stir casting method. *Journal of Materials Processing Technology* 1999; 119(1–3): 329–335. doi:10.1016/S0924-0136(01)00919-0.

16. Hashim J, Looney L and Hashmi MSJ. Metal matrix composites: Production by stir casting method, *Journal of Material Processing Technology*, 1999;92–93: 1–7.

17. Bharath V, Nagaral M, Auradi V and Kori SA. Preparation of 6061Al-Al2O3 MMC's by stir casting and evaluation of mechanical and wear properties. *Proceedia Material Science* 2014; 6 : 1658–1667.

18. Gopalakrishnan S and Murugan N. Production and wear characterisation of AA6061 matrix titanium carbide particulate reinforced composite by enhanced stir casting method. *Composite: Part B* 2012; 43: 302–308.

19. Boopathi M, Arulshri KP and Iyandurai N. Evaluation of mechanical properties of aluminium alloy 2024 reinforced with silicon carbide and fly ash hybrid metal matrix composites. *American Journal of Applied Sciences* 2013; 10 (3): 219–229.

20. Siva Prasad D and Shoba C. Hybrid composites – a better choice for wear resistant materials. *Journal of Materials Research and Technology* 2014; 3 (2): 172–178.

21. Chenghao L, Shusen W, Naibao H, Zhihong Z, Shuchun Z and Jing R. Effects of lanthanum and cerium mixed rare earth metal on abrasion and corrosion resistance of AM60 magnesium alloy. *Rare Metal Materials and Engineering* 2015; 44: 521–526.

22. Morgan PED, Marshall DB and Housley RM. High-temperature stability of monazite-alumina composites. *Materials Science and Engineering A* 1995; 195:215–222.

23. Gong G, Zhang B, Zhang H and Li W. Pressure less sintering of machinable Al2O3/LaPO4 composites in N2 atmosphere. *International Journal of Ceramics International* 2006; 32:349–352.

24. Abdul Majeeda M, Vijayaraghavana L, Malhotrab SK and Krishnamurthya R. Ultrasonic machining of Al2O3/LaPO4 composites. *International Journal of Machine Tools & Manufacture* 2008; 48:40–46.

25. Alangi N, Mukherjee J, Anupama P, Verma MK, Chakravarthy Y, Padmanabhan PVA, Das AK and Gantayet LM. Liquid uranium corrosion studies of protective yttria coatings on tantalum substrate. *Journal of Nuclear Materials* 2011; 410: 39–45.

26. Lloyd IK. Advances in Electroceramic Materials II: Ceramic Transaction volume 221 edited by Nair KM and Priya S (2010), pp. 115–124. ISBN: 978-0-470-92716-8. John Wiley and Sons inc, Hoboken, NJ.

27. Yihua H, Dongliang J, Jingxian Z and Qingling L. Fabrication of transparent lanthanum-doped yttria ceramics by combination of two-step sintering and vacuum sintering. Journal of American Ceramic Society 2009; 92(12): 2883–2887.

28. Sahin O, Demirkol I, Gocmez H, Tuncer M, Ali Cetinkerf angler H, SalihGuder H, Sahin E and RizaTuncdemir A. Mechanical properties of nanocrystalline tetragonal zirconia stabilized with CaO, MgO and Y2O3. *Proceedings of the 2nd International Congress APMAS2012*, April 26–29, 2012, Antalya, Turkey, Acta Physilicon carbidea Polonica A, Vol. 123 (2013).

29. Maiti K and Sil A. Preparation of rare earth oxide doped alumina ceramics, their hardness and fracture toughness determinations. *Indian Journal of Engineering and Material Science* 2006; 13: 443–450.

30. Cree D, Pugh M. Production and characterization of a three dimensional cellular metal-filled ceramic composite. *Journal of Material Processing Technology* 2010; 210: 1905–1917.

31. Lindroos VK and Talvitie MJ. Recent advances in metal matrix composites. *Journal of Material Processing Technology* 1995; 53: 273–284.
32. Tham LM, Gupta M and Cheng L. Effect of limited matrix-reinforcement interfacial reaction on enhancing the mechanical properties of aluminium-silicon carbide composites. *Acta Materialia* 2001; 49(16): 3243–3253. doi:10.1016/S1359-6454(01)00221-X.
33. Ray S. Synthesis of cast metal matrix particulate composites. *Journal of Materials Science* 1993; 28:5397–5413.
34. Abdulhaqq AH, Ghosh PK, Jain SC, Ray S. Processing, microstructure, and mechanical properties of cast in-situ Al (Mg, Mn) -Al2O3 (MnO2) composite. *Metallurgical and Materials Transactions A* 2005; 36A:2211–2223.
35. Kopac J and Krajnik P. Robust design of flank milling parameters based on grey-taguchi method. *Journal of Materials Processing Technology* 2007; 191:400–403.
36. Azmir MA and Ahsan AK. Investigation on glass/epoxy composite surfaces machined by abrasive waterjet machining. *Journal of Materials Processing Technology* 2008; 198: 122–128.
37. Sankerf angler S and Warrier KGK. Aqueous sol-gel synthesis of LaPO4 nano rods starting from lanthanum chloride precursor. *Journal of Sol-Gel Technology* 2011; 58:195–200.
38. Yusof AM and Ismail S. Multiple regression analysis as a tool to property investment decision making. *Proceedings of the 17th IBIMA Conference on Creating Global Competitive Economies: A 360-Degree Approach*, Soliman KS, Ed., vol. 1–4, pp. 700–708.
39. Ma JS. Grey target decision method for a variable target centre based on the decision maker's preferences. *Journal of Applied Mathematics* 2014; 2014: 6.
40. Liu SF, Cai H, Yang YJ and Cao Y. Advance in grey incidence analysis modeling. *Systems Engineering-Theory and Practice* 2013; 33(8):2041–2046.
41. Sun Y-G and Dang Y-G. Improvement on grey T's correlation degree. *Journal of System Science and Information* 2008; 4(4): 135–139.
42. Jeyapaul R and Shahabudeen P. Quality management research by multi response problem in the Taguchi method: A review. *Journal of Advanced Manufacturing Technology* 2005; 26:1331–1337.
43. Liu SF, Hu ML and Yang YJ. Progress of grey system models. *Transactions of Nanjing University of Aeronautics and Astronautics* 2012; 29(2): 103–111.
44. Gupta K, Jain NK and Laubscher R. *Chapter 4-Advances in Gear Manufacturing. Advanced Gear Manufacturing and Finishing*, Elsevier, 2017, 67–125.
45. Zitoune R and Bougherara H. Machining and drilling processes in composites manufacture: Damage and material integrity. *Advances in Composites Manufacturing and Process Design*, 2015, 177–195. Woodhead Publishing.
46. Hashish M. Optimization factors in abrasive-waterjet machining. *Journal of Manufacturing Science and Engineering* 2008; 113: 29.
47. Taylor R. Thermal conductivity. *Concise Encyclopedia of Advanced Ceramic Materials* 1991: 472–475. doi: 10.1016/B978-0-08-034720-2.50130-1.
48. Bishui B and Prasad J. Thermal conductivity of ceramic materials. *Transactions of the Indian Ceramic Society* 2014; 17: 108–116. doi: 10.1080/0371750X.1958.10855379.
49. Noorul Haq A, Marimuthu P and Jeyapaul R. Multi response optimization of machining parameters of drilling Al/Silicon carbide metal matrix composite using grey relational analysis in the Taguchi method. *Journal of Advance Manufacturing Technology* 2008; 37:250–255.
50. Balamurugan K, Uthayakumar M, Sankar S, Hareesh US and Warrier KGK. Mathematical modeling on multiple variables in machining LaPO4/Y2O3 composite by abrasive waterjet. *International Journal of Machining and Machiniability of Materials* 2017; 19(5): 426–439.

51. Balamurugan K, Uthayakumar M, Sankar S, Hareesh US and Warrier KGK. Preparation, characterization and machining of LaPO4-Y2O3 composite by abrasive water jet machine, *International Journal of Computer Aided Engineering and Technology* 2018; 10(6): 684–697.

52. Balamurugan K, Uthayakumar M, Sankar S, Hareesh US and Warrier KGK. effect of abrasive waterjet machining on LaPO4 /Y2O3 ceramic matrix composite. *International Journal of Australian Ceramic Society* 2018; 54(2): 205–214, doi: 10.1007/s41779-017-0142-7.

4 Drilling of Hybrid MMCs Using DLC- and HC-Coated Tools

T. Sampath Kumar
Vellore Institute of Technology University

M. Vignesh
Amrita Viswavidhyapeetam

A. Vinoth Jebaraj, P. Dilip Kumar,
N.V.S.S.S.K. Manne Dilip, and Abhishek Singh
Vellore Institute of Technology University

CONTENTS

DOI: 10.1201/9781003345466-4

4.1 INTRODUCTION

Current years have seen the involvement of cutting tool industry in developing new and harder coatings and cutting tool technology. In the past 50 years, hard-coated cutting tools played a major part of machining industry and now the demand for the development is increasing even more. The physical vapor deposition (PVD) and chemical vapor deposition (CVD) processes are successfully used for coating for a very long time. Nanostructured PVD coatings are the recent advancements and have been established very quickly. Reason behind using coated tools is the increased machinability and to get better results like better surface and lesser tool wear out of operation. TiAlN, TiO_2, Al_2O_3, TiN, TiC, AlCrN, TiCN and AlTiN are some of the commonly used conventional coatings [1]. These coatings are of great advantage as these not only provide us with better surface finish, better material removal rate (MRR), chip formation etc., but also with the high hardness and toughness, better thermal stability and better strength under extreme machining conditions. A countable number of researches have been done on hybrid Al-MMC with nanoparticles as reinforcements [2].

The tools basically used in the machining of magnesium, aluminum, titanium, chromium, etc., alloys as well as stainless steel are TiN-, CrN- and ZrN-coated carbide tools [3]. Out of all the cutting tools used for machining, ZrN outperformed compared to TiN- aand CrN-. Despite its irregular surface finish, it also led to good performances under heavy machining parameters [4]. The TiAlN-coated drill bit shows the suitability of drilling test without cutting fluid. TiN- and CrAlN-coated drill bits have suffered significantly with higher wear and lower level of performance. TiN coating gives higher life and faster penetration, but it is very expensive and the risk of damage of drill bit before its complete wear is always present. The properties of Ti- and Cr-coated high-speed steel Co5 and the drill bits made up of given steel are studied. The parameters taken into consideration are thickness, wear rate, hardness and volume of material loss [5]. The maximum hardness value achieved was 34.9 GPa for TiAlN coating. Coating characteristics are crucial for quality of machined product but the parametric machining condition also plays a vital role. Therefore, the selection of machining conditions is significant and should be taken care. MMCs are considered high strength with light-weight, high stiffness and are very much attracted by various industries such as automobile, aerospace and defense [6]. The machining parameters of Al-MMC in drilling operation was studied, and it was concluded that increasing the cutting speed and lowering the depth of cut and feed rate will give us the better surface finish. The inferences are drawn based on the response surface methodology (RSM). The workpiece used was aluminum metal matrix composite (Al-MMC) which is ductile in nature. The aluminum matrix is being reinforced with boron nitride (BN) and aluminum oxide (Al_2O_3) nanoparticles. The metal matrix composites (MMCs) are given a special attention for the last couple of years due to their high strength, excellent hardness, high wear resistance and less weight [7].

A lot of investigation has been done using different coating materials and using different machining materials. Coating materials like TiN, TiC and TiCN are very widely used in the machining industry because of their incredible machinability characteristics [8]. But the diamond is among the later not yet properly explored with coating. This research article will focus on using diamond-like carbon (DLC) coating and hard carbon (HC) coating for drilling tools. A hybrid MMC needs a special attention because of its unique properties. Here, BN and Al_2O_3 nanoparticles were reinforced with aluminum metal used as workpiece due to its less weight, high hardness and high strength. It is highly used in aeronautical, automobile as well as in medicine industry. This research mainly focuses on developing a diamond-based coating which can be as good as conventional coatings used and also banish few disadvantages. Different tests have been run to get the optimum results. The output parameters such as tool wear, circularity, surface roughness, chip formation analysis, cutting force were studied. The output parameters such as tool wear will be calculated by image profiler and surface roughness being examined by surface roughness tester, cutting force by Kistler force tool dynamometer and finally the comparison of the experimental results will be done. The suitability of the PVD-coated drill bits on machining Al-MMC was investigated. The use of DLC-coated and HC-coated tools has not been done yet. Hence in the present article, a detailed focus is on study of aluminum based metal matrix composites reinforced with BN and Al_2O_3 and the machinability investigation on fabricated composite material with DLC coating, HC coating and uncoated drill bits. The output parameters such as chip formation, cutting force, surface roughness and tool wear were analyzed, and the results are obtained by implementing Taguchi method approach. The study will help to get a fine machining tool coated either with DLC or HC coatings for machining of composite materials. The coated tool is having high strength, good surface finish, less tool wear and an overall better machinability. Also the coating has some extensive properties such as low friction, resistance to wear, chemical inertness etc. The experimental trial was conducted by drilling a through hole of 8 mm diameter and measuring the various output responses such as cutting force, surface roughness, MRR, hole circularity, chip analysis and taper ratio were recorded and analyzed.

4.2 MATERIAL AND METHOD

4.2.1 STIR CASTING METHOD TO FABRICATE HYBRID AL-MMC

Hybrid Al-MMC reinforced with nanoparticles of 2% of Al_2O_3 and 1% of BN was used as a workpiece material. Aluminum alloy powder (Al 7075) of 800 g was taken in a crucible, kept in a muffle electrical furnace and melted and the nanoparticles of Al_2O_3 and BN were mixed using a stir casting technique. The workpiece is formed by the casting process by pouring liquid metal into a mold of desired shape and size and then the liquid metal is allowed to solidify. Subsequently the metal is cooled and pulled out of the mold, and excess material is removed. Irregularities resulting from the imperfections of mold are very common in the casting process. The excess materials can be corrected by processes like cutting, grinding etc. The casted Al-MMC was machined to obtain the final size of $120 \times 100 \times 10$ mm workpiece in order to fix the workpiece properly in a specially designed fixture and to obtain precise results and to reduce the vibrations during the machining process.

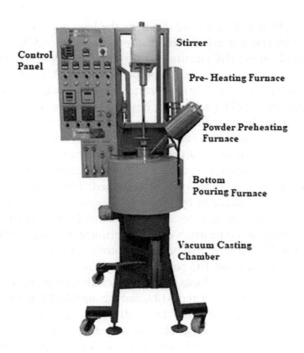

FIGURE 4.1   Stir casting equipment.

The hardness of the Al-MMC reinforced with nanoparticles of BN and Al_2O_3 was found to be 53.4 HRC. The scanning electron microscope (SEM) images of the workpiece are shown in Figure 4.2a and b with 500× and 1500× magnification scales, respectively. The elemental composition of the Al-MMC is shown in Figure 4.2c, and the EDAX of the workpiece is shown in Figure 4.2d. The microstructure of the Al-MMC shown in Figure 4.2a and b represents that the reinforcing particles are uniformly distributed in the matrix material. The elements present in the Al-MMC are confirmed with the elemental composition present in atomic % as shown in Figure 4.2c. The EDAX graph confirms the reinforcing of BN and Al_2O_3 particles present in the Al matrix.

4.2.2   Tool and Coating Material

The carbide drill bit was used as cutting tools to conduct the experiments. The DLC and HC coatings were deposited on the tungsten carbide drill bits. The uncoated and coated drill bits were shown in Figure 4.3a–c. The specifications of the drill bit are diameter = 8 mm, helix angle = 30°, shank diameter = 8 mm, cutting length = 40 mm, overall length of the drill bit = 81 mm, number of flute = 2, point angle = 118°. The cutting edge is longer, if the point angle is lesser.

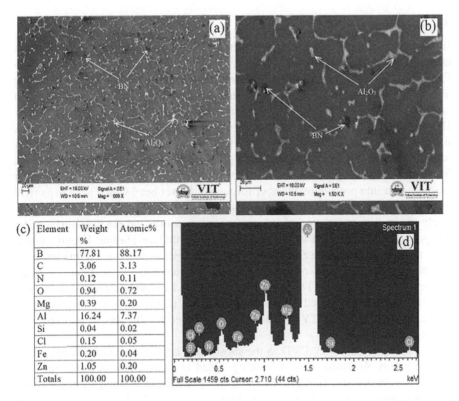

(c)

Element	Weight %	Atomic%
B	77.81	88.17
C	3.06	3.13
N	0.12	0.11
O	0.94	0.72
Mg	0.39	0.20
Al	16.24	7.37
Si	0.04	0.02
Cl	0.15	0.05
Fe	0.20	0.04
Zn	1.05	0.20
Totals	100.00	100.00

FIGURE 4.2 Microstructure of the Al-MMC workpiece. (a) SEM image of Al-MMC with magnification 500×. (b) SEM image of Al-MMC with magnification 1500×. (c) Elemental composition of Al-MMC. (d) EDAX of Al-MMC.

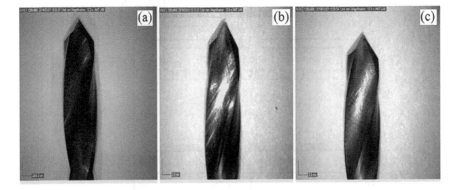

FIGURE 4.3 Carbide drill bits used for the experiments: (a) uncoated, (b) DLC coated, and (c) HC coated.

4.2.3 DLC COATING PROCESS

In the present research, DLC coating has been done on the tungsten carbide drill bits using the plasma-assisted chemical vapor deposition (PACVD) process as shown in Figure 4.4. A thin layer deposition is done from gas phase to a solid phase substrate and it involves a lot of chemical reactions. A suitable reacting gas is being taken and plasma is created. The creation of plasma takes place by running either AC or DC current between the two electrodes; the space between the cathode and anode is filled with a particular reacting gas. Plasma is the state of matter with a significant amount of ionized atoms or molecules of gas. The process uses the energy provided by plasma for the deposition with a temperature range of 200°C–400°C. Reactant gases stream into procedure chamber through a shower head which is an expansive punctured metal plate situated over the substrate. The shower head gives a progressively uniform appropriation of reactant gas stream over the reactant surface. An RF potential is connected to the shower head to produce plasma. Highly chemically reactive radicals are generated by high-speed electrons in the plasma ionize or separate reactant gases. Slender film of affidavit material above the sample is made by the resultant radicals. The high energy provided by plasma results as the key preferred merit of decreased temperatures for PACVD than the low pressure chemical vapor deposition (LPCVD). The energy provided by the plasma is transferred thermally [9].

The DLC is done by PACVD as the substrate is kept in a vacuum chamber containing argon. Acetylene (C_2H_2) is being added to the vacuum chamber pressure (2.5×10^{-3} mbar) with a flow rate of 100 sccm. The plasma is being ignited through the application of an electrical voltage (bias voltage of 500 V). Acetylene is broken down in fragments which contain hydrogen and carbon atoms. This results in the formation of an amorphous layer of diamond-like and carbon-like bonding components

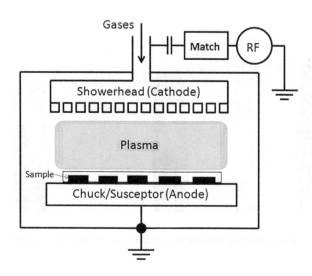

FIGURE 4.4 PACVD apparatus.

of hydrogenated DLC films (a-C:H) and hence the formation of a low friction hard DLC layer. The process is done at a temperature less than 250°C. The coating has an amorphous carbon nature and has 70% sp³ bonding with good protection against adhesive wear. The DLC coatings have low coefficient of friction (0.1–0.2) and suitable for dry running conditions. The DLC coating has high hardness of 20 GPa and coating thickness of 3 μm.

4.2.4 Hard Carbon Coating

PVD using cathode arc deposition process is as shown in Figure 4.5. In this process, using an electric arc material is vaporized from a cathode. This vaporized material then forms a thin film by condensing on the substrate. High current and low voltage, when strikes on cathode, result in the formation of high energy producing area known as cathode spot. A high-velocity stream of cathode material is formed as a result of the high concentrated temperature of the cathode spot and it leaves a cavity on the cathode. The cathode spot is active for a limited time, then it self-destroys and re-ignition is done near the previous cathode spot. The motion of arc is the result of this kind of behavior of cathode. As the circular segment is essentially a current conveying conductor, it very well may be affected by the use of an electromagnetic field, which practically speaking is utilized to quickly move the curve over the whole surface of the material, with the goal that the whole surface is disintegrated after some time. The high power density of arc is very effective in creating high level of ionization, differently charged ions, and neutral particles. The launching of the reactive gas in midst of the evaporation process will result in ionization and excitation.

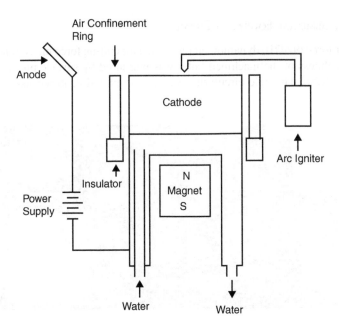

FIGURE 4.5 Cathodic arc process.

One drawback of the process is that if the cathode spot remains at an evaporative point for a really long time it can launch a lot of full scale particles or droplets. These are hindering to the execution of the coating as they are inadequately followed and can stretch out through the covering. More awful still if the cathode target material has a low melting point. The cathode spot can vanish through the substrate bringing about either the evaporation of material being dissipated or cooling water entering the chamber [10].

Short cylindrical conductive targets are being used, and the vacuum chamber is used as anode. Arc spots here are generated mechanically by short circuiting between anode and the cathode. Magnetic field is used to guide or steer the arc spots or can also move in random sense. The plasma beam consists of large groups of atoms or molecules. Here the cathode arc is made either in tubular or rectangular shape. HC coating is done using the steered arc cathodic deposition process. Tetrahedral amorphous carbon is the coating material used here. The deposition process maintains the current of 80 A, the substrate temperature of 450°C and a voltage of 200 volts. The coating has obtained a hardness of 40 GPa and a coating thickness of 3 μm. HC coating has low coefficient of friction, high thermal stability, protection against adhesion wear and better chip flow.

4.3   CNC VERTICAL MACHINE CENTER

The CNC Milling Machine (Make: Surya, model: VF 30 CNC VS as shown in Figure 4.6) was used to conduct the experimental trials. The drilling operation was performed in z-direction (feed direction).

4.3.1   EXPERIMENTAL EQUIPMENT DETAILS

The Kistler force (9247B) dynamometer was used for cutting force measurement. The surface roughness of the machined holes was measured by surface profiler (Make: Mahrsurf XR20). The hole circularity was measured and captured using Dinolite

FIGURE 4.6   CNC Milling machine used to conduct the L9 experimental trails.

and DinoCapture 2.0 software. The live image of the machined hole was captured and the differences in hole diameters of drilling were compared.

4.3.2 CALCULATION OF MATERIAL REMOVAL RATE

The MRR is calculated on the basis of weight difference of the workpiece before and after the drilling process. The time taken for drilling operation was recorded.

$$MRR = (W_i - W_f)/t(g/\min) \tag{4.1}$$

where W_i = initial weight of workpiece (g), W_f = final weight of workpiece (g) and t = time taken for drilling process (min).

4.3.3 MACHINING FACTORS AND THEIR LEVELS

The input parameters such as feed rate, spindle speed and type of drill bit were used in the present study. The designs of experiments with three levels of input parameters are given in Table 4.1. Taguchi's orthogonal array (L9) was used for the experiments during CNC drilling operation as shown in Table 4.2.

TABLE 4.1
Input Parametric Levels

Parameters	Levels		
	I	II	III
Spindle speed (rpm)	1200	2400	3600
Feed rate (mm/min)	40	80	120
Type of drill bit	UC	DLC	HC

TABLE 4.2
L9 Orthogonal Array for Drilling Experiments

Exp No.	Spindle Speed (rpm)	Feed Rate (mm/min)	Type of Drill Bits
1	1200	40	UC
2	1200	80	DLC
3	1200	120	HC
4	2400	40	DLC
5	2400	80	HC
6	2400	120	UC
7	3600	40	HC
8	3600	80	UC
9	3600	120	DLC

4.3.4 Design of Experiments

The signal-to-noise (S/N) ratio responses graph and the ANOVA tables have been obtained by using the Minitab software which gives better optimization results for each output parameters against the input parameters. The experiment was conducted according to Taguchi's orthogonal array (L9) and S/N ratio values have been obtained for each output. The S/N ratio response graphs have been plotted, and they give the optimum level for each output. The ANOVA table gives the contribution of each input parameters on influencing the output response. The Taguchi method followed for the experiment is a powerful tool in designing of parameters [11]. The methodology used for the design of input parameters is very valuable as the parameters are discrete and qualitative. It acts as a systematic approach and provides an efficient, simple tool to optimize the cost, design and performance of the drill bits.

4.4 RESULTS AND DISCUSSION

4.4.1 Cutting Force Analysis

The cutting force is the force exerted on the workpiece by the drill bit while performing the drilling operation. The cutting forces obtained for the experimental trails were shown in Figure 4.7. The mean value of the cutting force along the z-direction was considered and the highest cutting force obtained at L2 (287.4 N), and the minimum cutting force was obtained at L7 (69.23 N). The increase in spindle speed and feed rate increases the cutting force of the composite material, and hence, the tool flank wear increases due to the higher heat generation of the tool material and softens the drill bit during drilling the hybrid Al-MMC. With the increase in the spindle speed, the cutting forces also increases and improves the smoothness of the drill. Therefore, the tool wear increases because of the abrasive nature of the reinforcement BN and

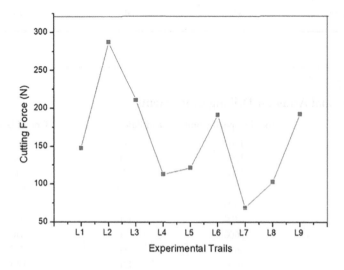

FIGURE 4.7   Cutting force graph obtained for L9 experimental trails.

Al_2O_3 nanoparticles. The built-up-edge (BUE) present on the drill bit increases the cutting force, and hence, the adhesive wear on the flank surface of the drill appears [12–15].

According to S/N ratios shown in Figure 4.8, the optimum level for the cutting force was found to be at spindle speed of 3600 rpm with a feed rate of 40 mm/min for high carbon coating. According to ANOVA results in Table 4.3, the cutting speed was found to be the most influencing parameter for cutting force by contributing 40.13% followed by feed rate as 34.07% and type of drill bit as 18.70%. HC coating was found to be the best tool to achieve lesser cutting force. The increase in cutting force will result in poor surface finish and lesser tool life; hence, spindle speed was found to be one of the most significant factors in deciding the cutting force generated

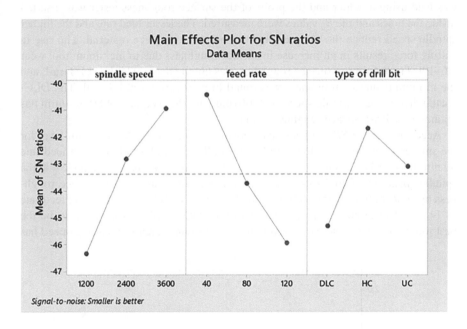

FIGURE 4.8   S/N ratio graph for cutting force.

TABLE 4.3

ANOVA Results for Cutting Force

Source	DF	Seq SS	Adj MS	F-Value	P-Value	Contribution (%)
Spindle speed	2	14,523	7262	5.65	0.150	40.13
Feed rate	2	12,331	6165	4.80	0.172	34.07
Type of drill	2	6,768	3384	2.63	0.275	18.70
Error	2	2,570	1285			7.10
Total	8	36,192				100.00

during drilling operation followed by the other factors. For the fabricated hybrid Al-MMC, the HC coating has obtained lesser cutting force followed by DLC coating and uncoated drill bits. The HC coating with a spindle speed of 3600 rpm and a feed rate of 40 mm/min has obtained the least cutting force which is suitable for machining Al-MMC.

4.4.2 Surface Roughness Analysis

The surface roughness is a measure of deviations on the surface from its reference level. Surface roughness will give the information of how smooth the machined surface looks. The Ra values (average surface roughness) were measured for all the drilled holes, and the surface roughness graph is shown in Figure 4.9. The workpiece was held using a holder and the probe of the surface roughness tester was inserted inside the holes and the R_a values were measured. The lesser feed rate and the higher spindle speed reduce the surface roughness of the workpiece material. The rise in cutting force results in an increase in surface roughness due to maximum tool wear [16–19]. The maximum surface roughness was obtained as 1.924 μm at L8 trail, and the minimum surface roughness was found to be 0.026 μm at L2 trail. The DLC-coated drill bit at a spindle speed of 1200 rpm with the feed rate of 80 mm/min has obtained the least surface roughness value.

According to the S/N ratios graph shown in Figure 4.10, the optimum level for the surface roughness was found to be at spindle speed of 1200 rpm with feed rate 80 mm/min for DLC-coated drill bit. According to ANOVA results in Table 4.4, the spindle speed was found to be the most influencing parameter for surface roughness by contributing 56.15% followed by the type of drill bit as 21.11% and feed rate as 14.11%. Hence out of the selected drill bits, DLC drill bit was found to be the best tool to achieve lesser surface roughness. Thus an increase in cutting speed has

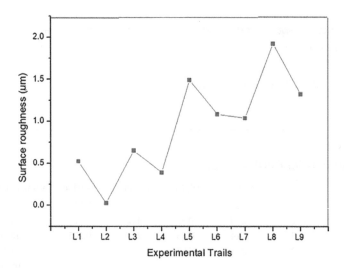

FIGURE.4.9 Surface roughness graph obtained for L9 experimental trails.

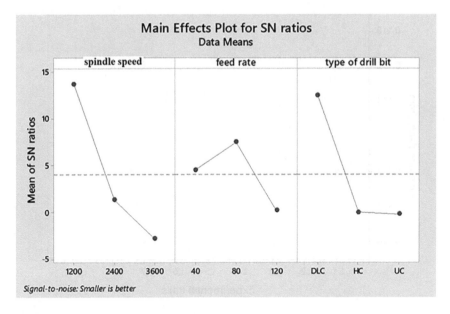

FIGURE 4.10   S/N ratio curve for surface roughness.

TABLE 4.4
ANOVA Results for Surface Roughness

Source	DF	Seq SS	Adj MS	F-Value	P-Value	Contribution (%)
Spindle speed	2	1.5987	0.7993	6.51	0.133	56.15
Feed rate	2	0.4019	0.2009	1.64	0.379	14.11
Type of drill	2	0.6010	0.3005	2.45	0.290	21.11
Error	2	0.2456	0.1228			8.63
Total	8	2.8471				100.00

more influence toward the surface roughness which causes poor surface finish on the machined surface of the workpiece (hybrid Al-MMC).

4.4.3   HOLE CIRCULARITY ANALYSIS

The circularity is a measure of diameter closeness of the drilled hole; it differs from that of diameter of the drill bit. By calculating the deviation in circularity, we can obtain which hole has the closest diameter to that of the drill bit size of 8 mm. The hole circularity images obtained for the L9 experimental trails are shown in Figure 4.11. The diameters of the circles have been measured with 30× magnification scale and shown in Figure 4.12. The higher feed rate and spindle speed has achieved better quality and performance of hole circularity with minimum deviation. The hole

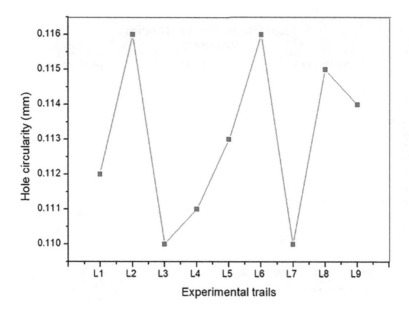

FIGURE 4.11   Hole circularity graph obtained for L9 experimental trails.

circularity increases, if the vibration of the drill bit increases during the machining process [20–23]. The least deviation in circularity was found in L3 and L7 trails which have obtained as 0.110 mm. The maximum deviation in circularity was found to be in L2 and L6 with a value of 0.116 mm. The lesser the deviation in circularity, more precise the drilled hole. So, the HC-coated drill bits were found to be with lesser deviation as per hole circularity analysis.

The deviation in hole circularity was analyzed against the input parameters like feed rate, cutting speed and type of drill bits from which the S/N graph and ANOVA table were obtained. According to the S/N ratios graph, the optimum level for the deviation in hole circularity was found to be at a spindle speed of 1200 rpm with feed rate 40 mm/min for HC-coated drill bit. According to the ANOVA results, the feed rate was the most contributing parameter for deviation in hole circularity by contributing 44.93% followed by type of drill bit as 40.58% and spindle speed as 1.45%. Hence out of the selected drill bits, HC-coated drill bit was found to be the best tool to achieve lesser deviation in hole circularity. The lesser deviation in hole circularity means more precise the size of the drilled hole and also better fit and surface finish; hence the deviation in hole circularity must be low for the better machining purposes (Figure 4.13 and Table 4.5).

4.4.4 MATERIAL REMOVAL RATE

MRR is a volume of the workpiece removed per unit time during the machining process. It is calculated experimentally by measuring the weight of the workpiece before and after each successive experimental trials of drilling operation. The higher the

FIGURE.4.12 Hole circularity images obtained for L9 experimental trails.

TABLE 4.5
ANOVA Results for Hole Circularity

Source	DF	Seq SS	Adj MS	F-Value	P-Value	Contribution (%)
Spindle speed	2	0.000001	0.000000	0.11	0.900	1.45
Feed rate	2	0.000021	0.000010	3.44	0.225	44.93
Type of drill	2	0.000019	0.000009	3.11	0.243	40.58
Error	2	0.000006	0.000003			13.04
Total	8	0.000046				100.00

MRR, lesser the time consumed for each drilling operation and the power consumed is also less during the drilling process. It can determine the overhead cost of the machining by including the electricity and labor costs. The cutting speed increases the cutting force, which improves the MRR and increases the cutting temperature

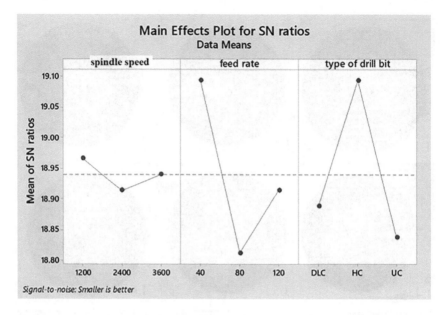

FIGURE 4.13   S/N ratio graph for hole circularity.

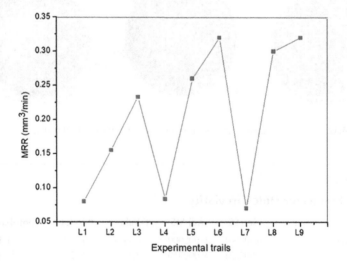

FIGURE 4.14   MRR graph obtained for L9 experimental trails.

during drilling due to abrasive wear on the tool. Hence the BUE formation occurs on the cutting tool edge and the flank wear increases [24–27]. The higher MRR was obtained at L9 trail as 0.32 mm³/min and the least MRR was obtained at L5 trail as 0.026 mm³/min. The DLC-coated drill bits has obtained higher MRR with spindle speed of 3600 rpm and feed rate of 120 mm/min (Figure 4.14).

The MRR was analyzed against the input parameters i.e. spindle speed, feed rate and type of drill bits from which the S/N graph and ANOVA tables were obtained as shown in Figure 4.15 and Table 4.6, respectively. According to the S/N ratios graph response, the optimum level for the MRR was found to be at spindle speed of 3600 rpm with feed rate 120 mm/min for DLC-coated drill bit. Also, according to the ANOVA results, the feed rate was found to be the most influencing parameter for MRR by contributing 81.85% followed by spindle speed of 11.14% and type of drill bit of 4.67%. Hence out of the selected drill bits, DLC-coated drill bit was found to be the best tool to achieve higher MRR. Higher MRR indicates lesser time for drilling operation with lesser power consumption.

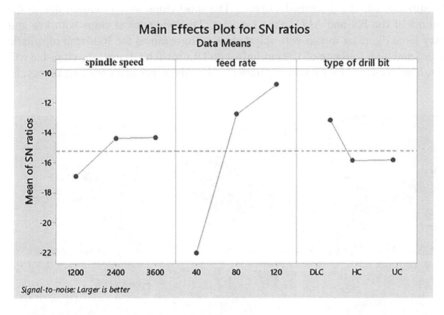

FIGURE 4.15   S/N ratio graph for MRR.

TABLE 4.6
ANOVA Result for MRR

Source	DF	Seq SS	Adj MS	F-Value	P-Value	Contribution (%)
Spindle speed	2	0.009946	0.004973	4.77	0.173	11.14
Feed rate	2	0.073045	0.036522	35.02	0.028	81.85
Type of drill	2	0.004168	0.002084	2.00	0.334	4.67
Error	2	0.002086	0.001043			2.34
Total	8	0.089245				100.00

4.4.5 CHIP ANALYSIS

The collection of chips in each successive experimental trial were obtained and analyzed. The different images of the chips obtained are shown in Figure 4.16. According to the observation of chips, the experimental trials of L4, L7 and L9 are found to be larger chip radius, which are easy to clean and dispose. When the feed rate raises, the curvature radius of the chip also increases. The long continuous conical chips were obtained for 2400 rpm of spindle speed with less feed rate of 40 mm/min with DLC drill bit. The continuous long conical with high and small diameters with flat radial chips were observed for 3600 rpm of spindle speed and low feed rate of 40 mm/min with HC drill bit. The long chips will damage the cutting drill bit surfaces and the machined surfaces. The short chips were produced due to the pullout of the BN and Al_2O_3 nano particles. The long conical chips with low and very large diameter mixed with spiral chips were obtained for 3600 rpm of spindle speed and high feed rate of 120 mm/min with DLC drill bit. Larger the chip size will result in lesser cost for cleaning operation and maintenance and are easier to recycle.

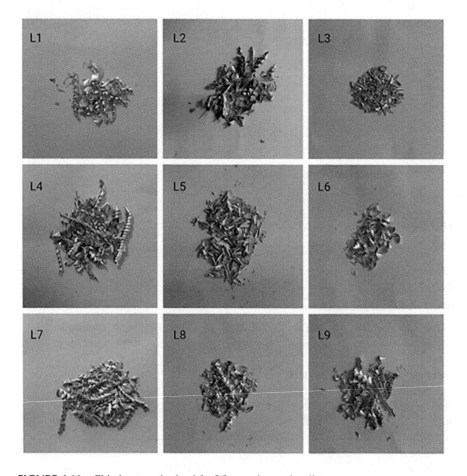

FIGURE 4.16   Chip images obtained for L9 experimental trails.

TABLE 4.7

Different Shapes of Chips Obtained for L9 Experimental Trials

Experimental Trials	Shapes of Chips Formed
L1	Short mixed chips with spiral type and rest with ribbon type
L2	Discontinuous with 3–4 turns ribbon type
L3	Tiny discontinuous conical chips with small radius
L4	Long continuous conical chips
L5	Tiny discontinuous conical chips mixed with 2–3 turns ribbon chips
L6	Thick flat radial chips with some conical chips
L7	Continuous long conical with high and small diameters with flat radial chips
L8	Discontinuous semi spiral chips mixed with continuous spiral chips
L9	Long conical chips with low and very large diameter mixed with spiral chips

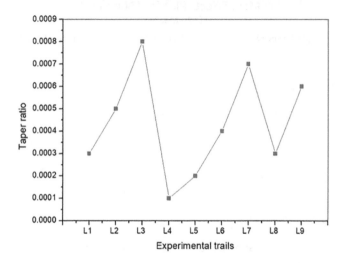

FIGURE 4.17 Taper ratio graph obtained for L9 experimental trails.

Due to the coating effect, no microchips were formed on the coated carbide tools. Discontinuous and short chips were easier for the removal of chips from the machining zone [28–30] (Table 4.7).

4.4.6 Taper Ratio Analysis

The taper ratio is the difference between the upper and lower diameter of the surface divided by the depth of the specimen. The taper ratio helps in determining the taper angle which is the measure of the deviation of the angle of the hole at the bottom compared to the top. The highest taper ratio was found at L3 trail with a value of 0.0008. And the least was found at L4 trail with the value of 0.0001. DLC-coated drill bit with spindle speed 2400 rpm and feed rate 40 mm/rev was found to be the best tool. Lesser the taper ratio, lesser the taper angle which results in straighter drilled holes [31,32] (Figure 4.17).

The taper ratio was analyzed against spindle speed, feed rate and type of drill bits from which the S/N ratio graph plots and ANOVA tables were recorded as shown Figure 4.18 and Table 4.8. According to S/N ratios graph response, the optimum level for the taper ratio was found to be at a spindle speed of 2400 rpm with a feed rate of 80 mm/min for diamond-like coating drill bit. Also, according to ANOVA results, the spindle speed was found to be the most influencing parameter by contributing 63.33% followed by feed rate as 21.11% and type of drill bit as 7.78%. The least optimum level for the taper ratio was found to be at 2400 rpm with 40 mm/min while drilling with DLC drill bit. Hence out of the selected drill bits, diamond-like coating drill bit was found to be the best tool to achieve lesser taper ratio. Taper ratio is used to determine the taper angle of the hole. Lesser the taper ratio, lesser the taper angle hence better fit and finish of the hole.

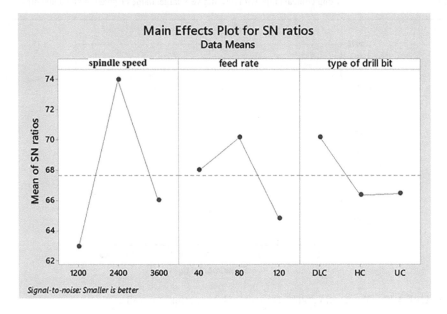

FIGURE 4.18   S/N ratio graph for tapered ratio.

TABLE 4.8
ANOVA Results for Tapered Ratio

Source	DF	Seq SS	Adj MS	F-Value	P-Value	Contribution (%)
Spindle speed	2	0.000001	0.000005	8.14	0.109	63.33
Feed rate	2	0.000003	0.000001	2.71	0.269	21.11
Type of drill	2	0.000002	0.000001	1.00	0.500	7.78
Error	2	0.000001	0.000002			7.78
Total	8	0.000001				100.00

4.5 CONCLUSIONS

Based on the drilling experimentation conducted on various coated tools, the following inferences are made:

- The optimum level for the cutting force was found to be at a spindle speed of 3600 rpm with feed rate of 40 mm/min with HC-coated drill bit and DLC-coated drill bit performs better for surface roughness criterion at a spindle speed of 1200 rpm with feed rate of 80 mm/min.
- The optimum level for the MRR was found at a spindle speed of 3600 rpm with feed rate of 120 mm/min with uncoated drill bit and for hole circularity HC-coated drill bit performs better with a spindle speed of 1200 rpm with feed rate 40 mm/min, respectively.
- For tapered ratio of drilled hole, the optimum level was found at a spindle speed of 2400 rpm with a feed rate of 80 mm/min using DLC-coated tool.
- The optimum level for obtaining the least cutting force, surface roughness, taper ratio, deviation in hole circularity and highest MRR was obtained at L5 trail with a spindle speed of 2400 rpm and a feed rate of 80 mm/min for HC-coated tool during machining of hybrid Al-MMC.

REFERENCES

1. Gariboldi, E. (2003). Drilling a magnesium alloy using PVD coated twist drills. *Journal of Materials Processing Technology*, 134(3), 287–295.
2. Kottfer, D., Ferdinandy, M., Kaczmarek, L., Maňková, I., & Beňo, J. (2013). Investigation of Ti and Cr based PVD coatings deposited onto HSS Co 5 twist drills. *Applied Surface Science*, 282, 770–776.
3. Haq, A. N., Marimuthu, P., & Jeyapaul, R. (2008). Multi response optimization of machining parameters of drilling Al/SiC metal matrix composite using grey relational analysis in the Taguchi method. *The International Journal of Advanced Manufacturing Technology*, 37(3), 250–255.
4. Rajmohan, T., Palanikumar, K., & Kathirvel, M. (2012). Optimization of machining parameters in drilling hybrid aluminium metal matrix composites. *Transactions of Nonferrous Metals Society of China*, 22(6), 1286–1297.
5. Kumar, J. P., & Packiaraj, P. (2012). Effect of drilling parameters on surface roughness, tool wear, material removal rate and hole diameter error in drilling of OHNS. *International Journal of Advanced Engineering Research and Studies*, 1(3), 150–154.
6. Barnes, S., Pashby, I. R., & Hashim, A. B. (1999). Effect of heat treatment on the drilling performance of aluminium/SiC MMC. *Applied Composite Materials*, 6(2), 121–138.
7. Sharma, P., & Sharma, R. International Journal of Latest Research in Science and Technology. ISSN (Online), 2278–5299.
8. Palanikumar, K., & Muniaraj, A. (2014). Experimental investigation and analysis of thrust force in drilling cast hybrid metal matrix (Al–15% SiC–4% graphite) composites. *Measurement*, 53, 240–250.
9. Folea, M., Roman, A., & Lupulescu, N. B. (2010). An overview of DLC coatings on cutting tools performance. *Academic Journal of Manufacturing Engineering*, 8(3), 30–36.
10. Fuchs, H., Keutel, K., Mecke, H., & Edelmann, C. (1999). Pulsed vacuum arc discharges on steered arc cathodes. *Surface and Coatings Technology*, 116, 963–968.

11. Kıvak, T., Samtaş, G., & Çiçek, A. (2012). Taguchi method based optimisation of drilling parameters in drilling of AISI 316 steel with PVD monolayer and multilayer coated HSS drills. *Measurement*, 45(6), 1547–1557.
12. Raj, A. M., Das, S. L., & Palanikumar, K. (2013). Influence of drill geometry on surface roughness in drilling of Al/SiC/Gr hybrid metal matrix composite. *Indian Journal of Science and Technology*, 6(7), 5002–5007.
13. Sreenivasulu, R. (2015). Optimization of burr size, surface roughness and circularity deviation during drilling of Al 6061 using Taguchi design method and artificial neural network. *Independent Journal of Management & Production*, 6(1), 93–108.
14. Ramulu, M., Rao, P. N., & Kao, H. (2002). Drilling of (Al2O3) p/6061 metal matrix composites. *Journal of Materials Processing Technology*, 124(1–2), 244–254.
15. Kivak, T., Habali, K., & ŞEKER, U. (2012). The effect of cutting paramaters on the hole quality and tool wear during the drilling of Inconel 718. *Gazi University Journal of Science*, 25(2), 533–540.
16. Balaji, M., Rao, K. V., Rao, N. M., & Murthy, B. S. N. (2018). Optimization of drilling parameters for drilling of TI-6Al-4V based on surface roughness, flank wear and drill vibration. *Measurement*, 114, 332–339.
17. Bobzin, K. (2017). High-performance coatings for cutting tools. *CIRP Journal of Manufacturing Science and Technology*, 18, 1–9.
18. Jadhav, S. S., Kakde, A. S., Patil, N. G., & Sankpal, J. B. (2018). Effect of cutting parameters, point angle and reinforcement percentage on surface finish, in drilling of AL6061/Al2O3p MMC. *Procedia Manufacturing*, 20, 2–11.
19. Folea, M., Roman, A., & Lupulescu, N. B. (2010). An overview of DLC coatings on cutting tools performance. *Academic Journal of Manufacturing Engineering*, 8(3), 30–36.
20. Taşkesen, A., & Kütükde, K. (2014). Experimental investigation and multi-objective analysis on drilling of boron carbide reinforced metal matrix composites using grey relational analysis. *Measurement*, 47, 321–330.
21. Chatterjee, S., Mahapatra, S. S., & Abhishek, K. (2016). Simulation and optimization of machining parameters in drilling of titanium alloys. *Simulation Modelling Practice and Theory*, 62, 31–48.
22. Balaji, M., Murthy, B. S. N., & Rao, N. M. (2016). Optimization of cutting parameters in drilling of AISI 304 stainless steel using Taguchi and ANOVA. *Procedia Technology*, 25, 1106–1113.
23. Nomani, J., Pramanik, A., Hilditch, T., & Littlefair, G. (2013). Machinability study of first generation duplex (2205), second generation duplex (2507) and austenite stainless steel during drilling process. *Wear*, 304(1–2), 20–28.
24. Assala, O., Fellah, M., Mechachti, S., Touhami, M. Z., Khettache, A., Bouzabata, B., & Jiang, X. (2018). Preparation and characterisation of diamond-like carbon films prepared by MW ECR/PACVD process deposited on 41Cr–Al–Mo7 nitrided steel. *Transactions of the IMF*, 96(3), 145–154.
25. Varade, A., Krishna, A., Reddy, K. N., Chellamalai, M., & Shashikumar, P. V. (2014). Diamond-like carbon coating made by RF plasma enhanced chemical vapour deposition for protective antireflective coatings on germanium. *Procedia Materials Science*, 5, 1015–1019.
26. Nakagawa, H., Kurita, Y., Ogawa, K., Sugiyama, Y., & Hasegawa, H. (2008). Experimental analysis of chatter vibration in end-milling using laser Doppler vibrometers. *International Journal of Automation Technology*, 2(6), 431–438.
27. Sahu, S. K., Ozdoganlar, O. B., DeVor, R. E., & Kapoor, S. G. (2003). Effect of groove-type chip breakers on twist drill performance. *International Journal of Machine Tools and Manufacture*, 43(6), 617–627.

28. Motorcu, A. R., Kuş, A., & Durgun, I. (2014). The evaluation of the effects of control factors on surface roughness in the drilling of Waspaloy superalloy. *Measurement*, 58, 394–408.

29. Kremer, A., Devillez, A., Dominiak, S., Dudzinski, D., & El Mansori, M. (2008). Machinability of AI/SiC particulate metal-matrix composites under dry conditions with CVD diamond-coated carbide tools. *Machining Science and Technology*, 12(2), 214–233.

30. Pendse, D. M., & Joshi, S. S. (2004). Modeling and optimization of machining process in discontinuously reinforced aluminium matrix composites. *Machining Science and Technology*, 8(1), 85–102.

31. Singh, S. (2016). Effect of modified drill point geometry on drilling quality characteristics of metal matrix composite (MMCs). *Journal of Mechanical Science and Technology*, 30(6), 2691–2698.

32. Tosun, G., & Muratoglu, M. (2004). The drilling of Al/SiCp metal–matrix composites. Part II: workpiece surface integrity. *Composites Science and Technology*, 64(10–11), 1413–1418.

5 Multi-Criteria Decision Making Approach for Selection of a Base Metal for Al-Based Metal Matrix Composites

Dileep Kumar Ganji, G. Rajyalakshmi,
Jayakrishna Kandasamy, and A. Deepa
Vellore Institute of Technology

G. Ranjith Kumar
Sri Venkateswara College of Engineering and Technology

CONTENTS

5.1   INTRODUCTION

Aluminium has an extensive application in general and industrial applications due to its varied attributes. Some of the attributes of Al are described herewith. The density of Al is one-third that of steel or copper; it is considered to be the lightest material and is resistant to weather. Al has high reflectivity and its strength is almost equivalent to that of construction steel. Al is highly elastic in nature and can be subjected to shock loads. Unlike carbon steels, Al maintains its toughness at very low temperatures and can easily be workable. Some of the shortfalls of aluminium as a base metal can be improved to a large extent by alloying Al with other metals or by reinforcing with other materials to make it a composite.

The major improvements in the properties of Al metal matrix composites include higher strength and stiffness, reduced density, enhanced high-temperature properties like thermal expansion coefficient, heat treatment, and better electrical performance [1].

For example, the elastic modulus of pure aluminium can be increased from 70 to 240 GPa by reinforcing with 60 vol% continuous alumina fibre. Also, the coefficient of thermal expansion can be decreased from 24 to 7 ppm/°C by addition of 60 vol % alumina fibre in pure aluminium [2].

It is a known fact that the aircraft manufacturers strictly rely on the life cycle approach in the material selection process as reduction of cost (optimization) is the most important criterion for them. This chapter focuses on choosing a best matrix material as Al alloy belonging to 2xxx and 7xxx series used for aircraft applications [3].

In Al metal matrix composites, Al as the matrix element for further reinforcement plays a vital role along with the reinforce material to obtain the required properties. Therefore, selection of an optimum matrix element is most important from the material designer's point of view in fabrication of the metal matrix composites [1].

5.2 AL ALLOY DESIGNATION AND APPLICATIONS

The alloys are designated using USA standard, which is almost adopted everywhere. Al alloys are light-weight, resistant to atmospheric corrosion, and exhibits good

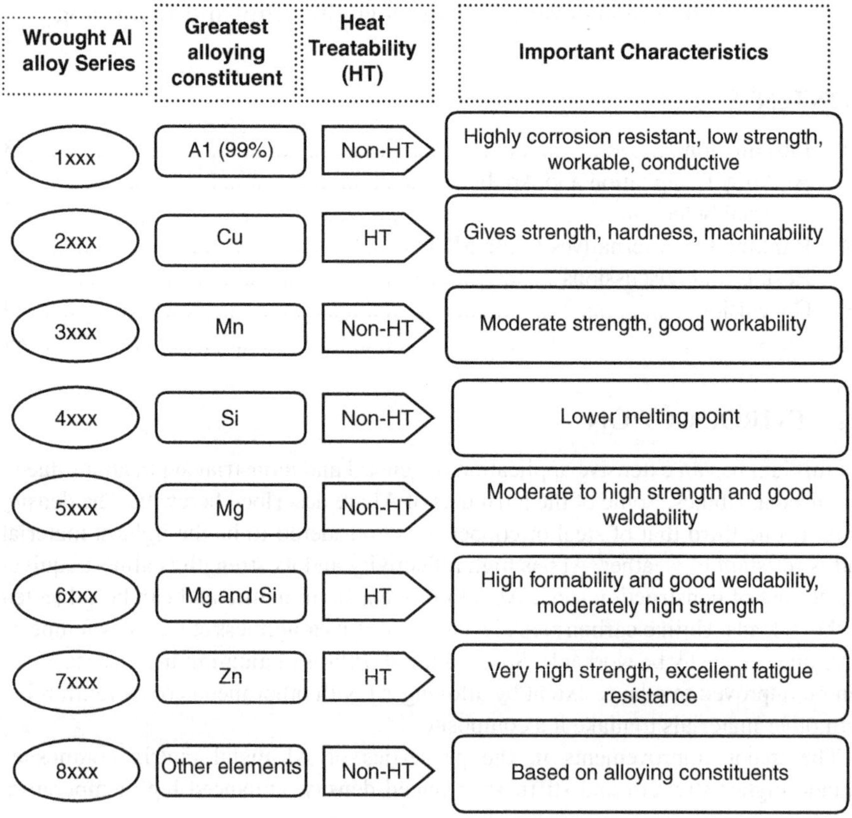

Wrought Al alloy Series	Greatest alloying constituent	Heat Treatability (HT)	Important Characteristics
1xxx	A1 (99%)	Non-HT	Highly corrosion resistant, low strength, workable, conductive
2xxx	Cu	HT	Gives strength, hardness, machinability
3xxx	Mn	Non-HT	Moderate strength, good workability
4xxx	Si	Non-HT	Lower melting point
5xxx	Mg	Non-HT	Moderate to high strength and good weldability
6xxx	Mg and Si	HT	High formability and good weldability, moderately high strength
7xxx	Zn	HT	Very high strength, excellent fatigue resistance
8xxx	Other elements	Non-HT	Based on alloying constituents

FIGURE 5.1 Al alloy designation and details.

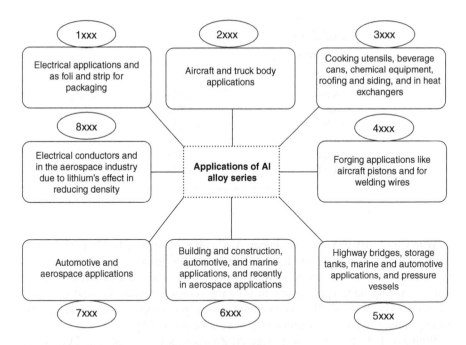

FIGURE 5.2 Applications of Al alloy configuration.

mechanical characteristics and can be used for wide applications in many fields. 1xxx alloys constitute almost 99% aluminium, whereas Al alloys majorly with Cu, Mn, SI, Mg, Mg and Si, Zn to form 2xxx, 3xxx, 4xxx, 5xxx, 6xxx and 7xxx series, respectively, in addition to other elements as minor constituents.

1xxx, 3xxx, 4xxx, 5xxx and 8xxx are non-heat-treatable alloys, whereas 2xxx, 6xxx and 7xxx are heat treatable. The characteristics of these alloys are depicted in Figure 5.1 [4–6].

With an objective to fly faster and farther, the aircraft industries are focusing on the materials that can withstand higher skin temperature with higher specific strengths. Al is still a most used material for aerospace applications and research is being focused on Al-Cu, AL-Zn and AL-Li series alloys [7]. Due to the environmental legislations imposed by many countries, fuel efficient vehicles are being developed with an objective to reduce weight by using Al to cast engine blocks, instead of cast iron. Cylinder heads, body frames, pistons, steering components and outer panels are some of the components where Al has its place [8,9]. The applications of wrought Al alloy configuration are presented in Figure 5.2 [4,5].

5.3 MATERIAL SELECTION

2xxx and 7xxx wrought aluminium alloy configuration is selected as the base metal for the aluminium metal matrix composite for the present study.

TABLE 5.1
Al Alloys

2xxx	7xxx
AA2011	AA7039
AA2014	AA7049
AA2024	AA7050
AA2048	AA7075
AA2219	AA7175
AA2618	AA7178

With Cu (2–10 wt%) as a constituent, Al-Cu alloys designated as 2xxx series with smaller composition of other elements can be precipitation hardened thus increasing strength with reduction in ductile characteristics. These are also heat treatable and are having huge applications in military vehicles, rocket fins and aerospace industries. The 2xxx Al alloy series do not possess good corrosion resistance as they undergo inter-granular corrosion under some conditions [3,5].

With Zn (1–8 wt%) as a constituent, Al-Zn alloys are designated as 7xxx series. By adding Zn, they can be heat treatable thereby enhancing strength through precipitation hardening. 7xxx series alloys are extensively used in aerospace for airframe structures, bicycle frames, baseball bats, armoured vehicles and other highly stressed parts. Generally Cu and Cr are also added in small quantities [3,5].

From the commercially available materials of 2xxx and 7xxx, the grades considered for analysis are presented in Table 5.1.

From the above discussion, it is evident that 2xxx series contains Cu as the major constituent and 7xxx contains Mg as the major constituent. 2xxx series exhibits higher strength, hardness and machinability characteristics whereas 7xxx series exhibit higher fatigue resistance in addition to better strength. The other alloying elements for these grades are Si, Fe, Cu, Mn, Mg, Cr, Ni, Zn, V, Zr, Ti and some unspecified elements of small proportions. The composition of various commercially available alloy materials considered for analysis is shown in Tables 5.2a and 5.2b [4].

Majority of the times, the manufacturers use trial-and-error experiments for identifying a suitable material for a specific application. Selection of a material using trial-and-error methods and without a scientific framework may lead to engineering failures and may lead to cost over runs and delay in the projects [10].

In the decision making approach of material selection, as a first step, based on the application, it is very much necessary to identify the basic characteristics and parameters of the material in accordance with design objectives. This methodology seems to be a multi-criteria decision making (MCDM), as different elements exhibit varied properties at different conditions.

TABLE 5.2A
Composition of 2xxx Al Alloy Configuration

2xxx Al Alloy	Si	Fe	Cu	Mn	Mg	Cr	Ni	Zn	V	Zr	Ti	Unspecified Other Elements		Al
												Each	Total	
												Weight %		
AA2011	0.40	0.70	5.00–6.00	-	-	-	-	0.30	-	-	-	0.05	0.15	Remaining
AA2014	0.50–1.20	0.70	3.90–5.00	0.40–1.20	0.20–0.80	0.10	-	0.25	-	-	0.15	0.05	0.15	Remaining
AA2024	0.50	0.50	3.80–4.90	0.30–0.90	1.20–1.80	0.10	-	0.25	-	-	0.15	0.05	0.15	Remaining
AA2048	0.15	0.20	2.80–3.80	0.20–0.60	1.20–1.80	-	-	0.25	-	-	010	0.05	0.15	Remaining
AA2219	0.20	0.30	5.80–6.80	0.20–0.70	0.02	-	-	0.10	0.05–015	0.10–0.25	0.02–0.10	0.05	0.15	Remaining
AA2618	0.10–0.25	0.90–1.30	1.90–2.70	-	1.30–1.80	-	0.90–1.20	0.10	-	-	0.04–0.10	0.05	0.15	Remaining

TABLE 5.2B
Composition of 7xxx Al Alloy Configuration

| 7xxx Al Alloy | Weight % | | | | | | | | | | Unspecified Other Elements | | |
	Si	Fe	Cu	Mn	Mg	Cr	Zn	Zr	Ti		Each	Total	Al
AA7039	0.30	0.40	0.10	0.10–0.40	2.30–3.30	0.15–0.25	3.50–4.50	-	0.10		0.05	0.15	Remaining
AA7049	0.25	0.35	1.20–1.90	0.20	2.00–2.90	0.10–0.22	7.20–8.20	-	0.10		0.05	0.15	Remaining
AA7050	0.12	0.15	2.00–2.60	0.10	1.90–2.60	0.04	5.70–6.70	0.08–0.15	0.06		0.05	0.15	Remaining
AA7075	0.40	0.50	1.20–2.00	0.30	2.10–2.90	0.18–0.28	5.10–6.10	-	0.20		0.05	0.15	Remaining
AA7175	0.15	0.20	1.20–2.00	0.10	2.10–2.90	0.18–0.28	5.10–6.10	-	0.10		0.05	0.15	Remaining
AA7178	0.40	0.50	1.60–2.40	0.30	2.40–3.10	0.18–0.28	6.30–7.30	-	0.20		0.05	0.15	Remaining

5.4 RANKING THE ALTERNATIVES USING MCDM

The selection of suitable elemental composition seems to be an MCDM problem with conflicting and diverse objectives, as different elements execute different properties at varied conditions [11]. The present study deals with the procedure that can be applied to identify suitable Al alloy among 2xxx series and 7xxx configuration for aerospace applications considering the key properties such as density, elastic modulus, yield and ultimate tensile strengths, percent elongation and fatigue endurance limit. The methodology deals with the selection from 'n' number of alternatives to give desired properties and applying MCDM algorithms to rank and select one among the commercially available candidate materials [12,13].

An ever-expanding progress in the mechanical advancements prompts improvement of a material having its own attributes, applications, impediments and preferences. Materials are liable for the viable usefulness, structure and strength of the product that assumes a significant role in the material plan and assembling process. Choice of material in building configuration process is troublesome because of tremendous number of unique materials accessibility and should consider qualitative and quantitative criteria [14]. The step-by-step approach used to rank and select the best alloy composition using MCDM techniques is shown in Figure 5.3.

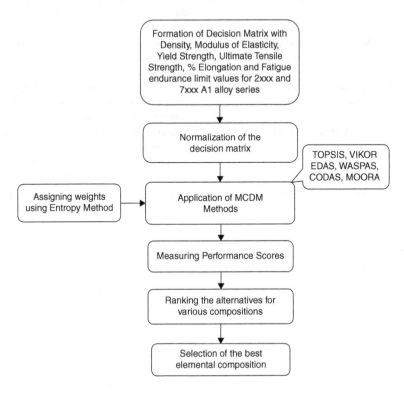

FIGURE 5.3 MCDM approach.

For the present work, the MCDM techniques explained in Table 5.3 are considered for ranking the alternatives and the corresponding computation methodology is taken from the literature [14–18]. The MCDM methods can be helpful in reducing the subjectivity in decision making by introducing a series of filters thus providing a choice among the complex alternatives. These are characterized by some specific mathematical relations and do not totally depend on the standards of judging among the alternatives, but consider multi-criteria for selection [19].

The criteria considered for ranking of alternate candidate materials are presented in Table 5.4 along with the units of measurement.

TABLE 5.3
MCDM Techniques

	Name of the MCDM Method	Objective
1.	TOPSIS (Technique for Order of Preference by Similarity to Ideal Solutions)	Assign ranks based on the shortest distance from ideal positive solution and farthest from ideal negative solution
2.	EDAS (Evaluation Based on Distance from Average Solution)	Assign ranks based on the positive and negative distances from the average solution
3.	WASPAS (Weighted Aggregated Sum Product Assessment)	Assign ranks based on the weighted sum of the normalized performance value and weighted product method value
4.	MOORA (Multi-objective Optimization on the Basis of Ratio Analysis)	Assign ranks to the problems with conflicting criteria by assigning ratios to the matrix of responses
5.	VIKOR (VlseKriterijumska Optimizacija I Kompromisno Resenje)	Assign ranks to the problems with conflicting criteria with a compromise ranking list which is closer to the ideal value
6.	CODAS (Combinative Distance-Based Assessment)	Assign ranks by using Euclidean distance and Taxicab distances as the primary and secondary measures respectively, and calculated through negative-ideal point

TABLE 5.4
Criteria of Material

Criteria	Unit
Density	g/cc
Yield Stress	MPa
Ultimate Tensile Stress	MPa
Elongation	%
Modulus of Elasticity	GPa
Fatigue Endurance Limit	MPa

The density of Al is one-third that of steel or copper and it is considered to be the lightest material and this allows better payloads for transportation vehicles due to high strength to weight ratio. From the strength point of view, pure Al does not have a high tensile strength. However, the addition of alloying elements like Mn, Cu, Si, Mg can increase the strength properties paving for tailored applications.

The decision matrix (properties of alternate candidate materials) with m available alternatives and n constraints is as follows and the decision matrix for the alternatives developed in the present study is provided in Tables 5.5a and 5.5b.

$$[X_{ij}]_{n\times m} = \begin{bmatrix} X_{11} & X_{12} & \cdots & X_{1m} \\ X_{21} & X_{22} & \cdots & X_{2m} \\ \vdots & \vdots & \vdots & \vdots \\ X_{n1} & X_{n2} & \cdots & X_{nm} \end{bmatrix}$$

TABLE 5.5A
Criteria Values of the Alternative Alloys of 2xxx Configuration (Decision Matrix)

| 2xxx Al Alloy | Temper | Density (g/cc) | Elastic Modulus (GPa) | Tensile Properties | | | Fatigue Endurance Limit (MPa) |
				Yield (MPa)	Tensile (MPa)	% Elongation	
AA2011	T3	2.83	70	295	380	15	125
AA2014	T6	2.8	72.4	415	485	13	125
AA2024	T4	2.77	72.4	325	470	20	140
AA2048	T8	2.75	70	416	457	13	110
AA2219	T62	2.84	73.8	290	415	10	105
AA2618	T61	2.76	74	370	440	10	125

TABLE 5.5B
Criteria Values of the Alternative Alloys of 7xxx Configuration (Decision Matrix)

| 7xxx Al Alloy | Temper | Density (g/cc) | Elastic Modulus (GPa) | Tensile Properties | | | Fatigue Endurance Limit (MPa) |
				Yield (MPa)	Tensile (MPa)	% Elongation	
AA7039	T6	2.73	69.6	380	450	13	125
AA7049	T73	2.82	72	450	515	12	115
AA7050	T74	2.83	70.3	450	510	13	170
AA7075	T6	2.8	71	505	570	11	160
AA7175	T736	2.8	72	476	490	14	159
AA7178	T76	2.83	71.7	440	572	17	130

TABLE 5.6

Procedure of Various MCDM Methods

	TOPSIS	VIKOR	WASPAS	MOORA	EDAS
Step 1	Define decision matrix, $[x_{ij}]_{n\times m}$ $= \begin{bmatrix} x_{11} & x_{12} & \cdots & x_{1M} \\ x_{21} & x_{22} & \cdots & x_{2m} \\ \vdots & \vdots & & \vdots \\ x_{n1} & x_{n2} & & x_{nm} \end{bmatrix}$	Define decision matrix, $[x_{ij}]_{n\times m}$ $= \begin{bmatrix} x_{11} & x_{12} & \cdots & x_{1M} \\ x_{21} & x_{22} & \cdots & x_{2m} \\ \vdots & \vdots & & \vdots \\ x_{n1} & x_{n2} & & x_{nm} \end{bmatrix}$	Define decision matrix, $[x_{ij}]_{n\times m}$ $= \begin{bmatrix} x_{11} & x_{12} & \cdots & x_{1M} \\ x_{21} & x_{22} & \cdots & x_{2m} \\ \vdots & \vdots & & \vdots \\ x_{n1} & x_{n2} & & x_{nm} \end{bmatrix}$	Define decision matrix, $[x_{ij}]_{n\times m}$ $= \begin{bmatrix} x_{11} & x_{12} & \cdots & x_{1M} \\ x_{21} & x_{22} & \cdots & x_{2m} \\ \vdots & \vdots & & \vdots \\ x_{n1} & x_{n2} & & x_{nm} \end{bmatrix}$	Define decision matrix, $[x_{ij}]_{n\times m}$ $= \begin{bmatrix} x_{11} & x_{12} & \cdots & x_{1M} \\ x_{21} & x_{22} & \cdots & x_{2m} \\ \vdots & \vdots & & \vdots \\ x_{n1} & x_{n2} & & x_{nm} \end{bmatrix}$
Step 2	Assign weights to all criterion	Assign weights to all criterion	Assign weights to all criterion	Assign weights to all criterion	Assign weights to all criterion
Step 3	Normalize the decision matrix using $\bar{X}_{ij} = \dfrac{X_{ij}}{\sqrt{\sum_{i=1}^{n} X_{ij}^2}}$	Find best $(x_{ij})_{max}$ for beneficial and $(x_{ij})_{min}$ for non-beneficial and best $(x_{ij})_{max}$ for beneficial and $(x_{ij})_{min}$ for non-beneficial	Normalize the decision matrix using $\bar{X}_{ij} = \dfrac{X_{ij}}{\sqrt{\sum_{i=1}^{n} X_{ij}^2}}$	Normalize the decision matrix using $\bar{X}_{ij} = \dfrac{X_{ij}}{\sqrt{\sum_{i=1}^{n} X_{ij}^2}}$	Determine average solution $AV_j = \dfrac{\sum_{i=1}^{n} X_{ij}}{n}$
Step 4	Weighted normalized matrix $V_{ij} = \bar{X}_{ij} W_j$	Calculate unity measure $S_i = \sum_{j=1}^{m} w_j \left[\dfrac{X_i^+ - X_{ij}}{X_i^+ - X_i^-} \right]$	Weighted normalized matrix $V_{ij} = \bar{X}_{ij} W_j$	Weighted normalized matrix $V_{ij} = \bar{X}_{ij} W_j$	Calculate positive distance average (PDA) and negative distance from the average (NDA) $PDA = [PDA_{ij}]_{n\times m}$ $NDA = [NDA_{ij}]_{n\times m}$
Step 5	Calculate ideal best (V_j^+) and ideal worst (V_j^-) for all criterion	Individual regret $R_j = \max_j \left[w_j \dfrac{X_i^+ - X_{ij}}{X_i^+ - X_i^-} \right]$	Compute preference value of weighted sum model $A^{WSM} = \sum_{i=1}^{n} W_j X_{ij}$	Calculate normalized assessment value (preferences score) $y_i = \sum_{j=1}^{g} w_j \bar{X}_{ij} - \sum_{j=g+1}^{n} w_j \bar{X}_{ij}$ where g is the number of beneficial criteria and $(n-g)$ is the number of non-beneficial criteria	If jth criterion is beneficial $PDA_{ij} = \dfrac{\max\left(0,(X_{ij} - AV_j)\right)}{AV_j}$ $NDA_{ij} = \dfrac{\max\left(0,(AV_j - X_{ij})\right)}{AV_j}$

EDAS (continued): Calculate negative ideal solution $ns = [ns_j]_{1 \times m}$, $ns_j = \min V_{ij}$. The Euclidean (E_i) and Taxicab (T_i) distances of alternatives from negative ideal solution is calculated using

(Continued)

TABLE 5.6 (Continued)
Procedure of Various MCDM Methods

	TOPSIS	VIKOR	WASPAS	MOORA	EDAS
Step 6	Calculate Euclidean distance from ideal best and ideal worst $$S_i^+ = \left[\sum_{j=1}^m (V_{ij} - V_j^+)^2\right]^{0.5}$$ $$S_i^- = \left[\sum_{j=1}^m (V_{ij} - V_j^-)^2\right]^{0.5}$$	Best values of S_i and R_i $$S^+ = \min_i S_i$$ $$R^+ = \min_i R_i$$ $$S^- = \max_i S_i$$ $$R^- = \max_i R_i$$	Compute preference value of weighted product model $$A_i^{WPM} = \prod_{j=1}^n X_{ij}^{w}$$	Give based on preferences value from higher to lower the least	If jth criterion is non-beneficial $$PDA_{ij} = \frac{\max\left(0, AV_j - X_{ij}\right)}{AV_j}$$ $$NDA_{ij} = \frac{\max\left(0, \left(X_{ij} - AV_j\right)\right)}{AV_j}$$ Calculate weighted sum of PDA and NDA $$SP_i = \sum_{j=1}^m w_j PDA_{ij}$$ $$SN_i = \sum_{j=1}^m w_j NDA_{ij}$$ Normalize the values of SP and SN $$NSP_i = \frac{SP_i}{\max_i SP_i}$$ $$NSN_i = 1 - \frac{SN_i}{\max_i SN_i}$$ The relative assessment is obtained using $$R_u = [h_{ik}]$$ $$h_{ik} = (E_i - E_k) + \Psi(E_i - E_k)(T_i - T_k)$$ Where Ψ is the threshold function to recognize equality of distances of two alternatives $$E_i = \sqrt{\sum_{j=1}^m (V_{ij} - ns_j)^2}$$ $$T_i = \sum_{j=1}^m |V_{ij} - ns_j|$$
Step 7	Calculate performance score $$P_i = \frac{S_i^-}{S_i^+ + S_i^-}$$	Calculate weight for the strategy for maximum group utility (performance score) $$Q_i = v\frac{S_i - S^+}{S^- - S^+} + (1-v)\frac{R_i - R^+}{R^- - R^+}$$	Compute preference value of WASPAS $$Q_i = \lambda A_i^{WSM} + (1-\lambda)A_i^{WPM}$$		The final appraisal score $$AS_i = \frac{1}{2}(NSP_i + NSN_i)$$ Final assessment score $$H_i = \sum_{k=1}^n h_{ik}$$
Step 8	Give ranks based on preference value from higher to lower the least	Give ranks based on preference value from lower to higher the least	Give ranks based on preference value from higher to lower the least		Give ranks based on preference value from higher to lower the least Give ranks based on preference value from higher to lower the least

As discussed, the various MCDM methods are considered for ranking the alternatives and best among these alternate candidate materials in the present study are TOPSIS (Technique for Order of Preference by Similarity to Ideal Solutions), EDAS (Evaluation Based on Distance from Average Solution), WASPAS (Weighted Aggregated Sum Product Assessment), MOORA (Multi-Objective Optimization on the Basis of Ratio Analysis), VIKOR (VlseKriterijumska Optimizacija I Kompromisno Resenje) and CODAS (Combinative Distance-Based Assessment). The step-by-step ranking procedure for the said methods is explained in Table 5.6 [14,19]. In general, the decision matrix for various alternatives with desired criteria is formed. These values are normalized by using linear or vector methods. The weights are assigned to all the criteria such that the sum of all the weights is equal to unity. Then the performance scores are calculated by applying the chosen techniques. These scores are ranked by arranging in either ascending or descending order. For TOPSIS, EDAS, MOORA and CODAS, higher performance is ranked first and lower score as the least, whereas for VIKOR method, lower performance score is ranked high and higher score is ranked low. These computations are done by using MATLAB.

5.5   RESULTS AND DISCUSSIONS

Entropy method is used to compute the criterion weights [20]. The first decision matrix is normalized using the below formula:

$$r_{ij} = \frac{x_{ij}}{\sum_{i=1}^{m} x_{ij}}$$

Then entropy is calculated using the relation:

$$e_j = -h \sum_{i=1}^{m} r_{ij} \ln(r_{ij}), j = 1,2,3,....,n, \quad \text{where} \quad h = \frac{1}{\ln(m)}, \text{ where 'm' is number of}$$

alternatives.

Afterwards, weight vector is computed using:

$$w_j = \frac{1 - e_j}{\sum_{j=1}^{n} (1 - e_j)}$$

The weights should be always in between 0 and 1. Also, the sum of all the weights should be equal to 1. The weights obtained are shown in Table 5.7, and these are assigned to the criterion during the ranking procedure after normalizing the values.

The performance scores are obtained for the MCDM methods employed (TOPSIS, VIKOR, WASPAS, MOORA, EDAS and CODAS) using the procedure mentioned in Table-6 and coding was done using MATLAB and the scores obtained are presented in Tables 5.8a and 5.8b.

For the performance scores obtained using the beneficial and non-beneficial criteria, the ranks are assigned to the alternatives developed and this is shown in Tables 5.9a and 5.9b.

TABLE 5.7

Weights Using Entropy Method

| Criteria | Density | Elastic Modulus | Tensile Properties | | | Fatigue Endurance Limit |
			Yield	Tensile	% Elongation	
Weights Assigned for 2xxx	0.0022	0.0059	0.2214	0.0661	0.6143	0.0901
Weights Assigned for 7xxx	0.0040	0.0042	0.1332	0.1268	0.3521	0.3797

TABLE 5.8A

Performance Scores for 2xxx Alloy

2xxx Alloy Series	TOPSIS	VIKOR	WASPAS	MOORA	EDAS	CODAS
AA2011	0.4860	0.0000	0.7569	0.4074	0.3332	0.2469
AA2014	0.3341	0.5000	0.7669	0.4084	0.6880	0.0106
AA2024	0.8865	0.0000	0.9471	0.5152	0.9412	1.3963
AA2048	0.3326	0.8896	0.7547	0.4023	0.5394	0.0066
AA2219	0.0115	1.0000	0.5876	0.3123	0.0166	−0.5406
AA2618	0.1065	0.8687	0.6393	0.3402	0.3804	−0.3632

TABLE 5.8B

Performance Scores for 2xxx Alloy

7xxx Alloy Series	TOPSIS	VIKOR	WASPAS	MOORA	EDAS	CODAS
AA7039	0.3010	0.0000	0.7874	0.2719	0.2287	−0.2027
AA7049	0.1499	1.0000	0.7379	0.2486	0.0159	−0.3518
AA7050	0.5903	0.0000	0.8795	0.3176	0.6291	0.2392
AA7075	0.4271	0.6145	0.8133	0.2850	0.2836	0.0662
AA7175	0.6078	0.0716	0.8672	0.3115	0.5317	0.1483
AA7178	0.6086	0.2458	0.8889	0.3249	0.8347	0.2506

The ranks assigned based upon the performance scores for 2xxx and 7xxx candidate materials are plotted and are shown in Figure 5.4. From the graph, for five of the methods, except VIKOR, AA7178 among 7xxx series is ranked first and AA2024 among 2xxx is ranked first by all the MCDM techniques applied. So, it is clear that, all the methods are giving almost same rank with minor deviations. But it is always preferable to perform the consistency check.

TABLE 5.9A
Ranks Assigned for 2xxx Al Alloys

2xxx Alloy Series	TOPSIS	VIKOR	WASPAS	MOORA	EDAS	CODAS
AA2011	2	2	3	3	5	2
AA2014	3	3	2	2	2	3
AA2024	1	1	1	1	1	1
AA2048	4	5	4	4	3	4
AA2219	6	6	6	6	6	6
AA2618	5	4	5	5	4	5

TABLE 5.9B
Ranks Assigned for 7xxx Al Alloys

7xxx Alloy Series	TOPSIS	VIKOR	WASPAS	MOORA	EDAS	CODAS
AA7039	5	2	6	5	5	5
AA7049	6	6	5	6	6	6
AA7050	3	1	2	2	2	2
AA7075	4	5	4	4	4	4
AA7175	2	3	3	3	3	3
AA7178	1	4	1	1	1	1

To check the consistency in the ranking and select the best from the alternatives developed, rank correlation study is performed using Spearman rank correlation coefficient [14]. This coefficient can be worked out using the following relationship to see the correlation between the various methodologies applied.

$$r_R = 1 - \frac{6 \sum_i d_i^2}{n(n^2 - 1)}$$

where n is the number of data points of the two variables and d_i is the difference in the ranks of the ith element of each random variable considered. The value of Spearman correlation coefficient ranges from +1 to −1. The rank correlation coefficient for the methods employed, i.e. TOPSIS, VIKOR, WASPAS, MOORA, EDAS and CODAS, is calculated and is shown in Tables 5.10a and 5.10b.

Neglecting the weak coefficient of correlation, it is evident that almost all the values are having a strong positive coefficient of correlation from 0.66 to 1.00, which implies that all the methods are closely related and the values obtained can be used

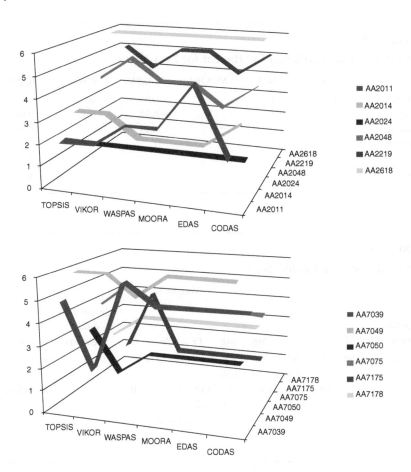

FIGURE 5.4   Ranking with different MCDM techniques.

TABLE 5.10A
Spearman Rank Correlation Coefficient for 2xxx Alloys

	CODAS	EDAS	MOORA	WASPAS	VIKOR	TOPSIS
CODAS	1	0.66	0.94	0.94	0.94	1
EDAS		1	0.83	0.83	0.6	0.66
MOORA			1	1	0.89	0.94
WASPAS				1	0.89	0.94
VIKOR					1	0.94
TOPSIS						1

for experimentation. So, the best alternative in 2xxx series is AA2024 and in 7xxx series is AA7178.

TABLE 5.10B
Spearman Rank Correlation Coefficient for 7xxx alloys

	CODAS	EDAS	MOORA	WASPAS	VIKOR	TOPSIS
CODAS	1	1	1	0.94	0.43	0.94
EDAS		1	1	0.94	0.43	0.94
MOORA			1	0.94	0.43	0.94
WASPAS				1	0.2	0.89
VIKOR					1	0.31
TOPSIS						1

TABLE 5.11
Selected Alternate Candidates

	Weight %									
Al as Base Element	Si	Fe	Cu	Mn	Mg	Cr	Zn	Ti	Unspecified Other Elements	Al
AA2024	0.50	0.50	3.80–4.90	0.30–0.90	1.20–1.80	0.10	0.25	0.15	0.20	Remaining
AA7178	0.40	0.50	1.60–2.40	0.30	2.40–3.10	0.18–0.28	6.30–7.30	0.20	1.20	Remaining

TABLE 5.12
Properties of the Best Alternate Materials

Al Alloy	Temper	Density (g/cc)	Elastic Modulus (GPa)	Tensile Properties			Fatigue Endurance Limit (MPa)
				Yield (MPa)	Tensile (MPa)	% Elongation	
AA2024	T4	2.77	72.4	325	470	20	140
AA7178	T76	2.83	71.7	440	572	17	130

The best alternatives to proceed for experimentation obtained through ranking are shown in Table 5.11 based upon the criteria considered.

The properties of the selected alternate materials of commercially available 2xxx and 7xxx Al alloy materials are shown in Table 5.12. When these are used as matrix material (base metal) in metal matrix composites, these values will further be enhanced and can be suited for higher applications where the components are subjected to higher stresses.

5.6   CONCLUSION

The ranking of the alternatives of 2xxx (AA2011, AA2014, AA2024, AA2048, AA2219, AA2618) and 7xxx (AA7039, AA7049, AA7050, AA7075, AA7175, AA7178) commercially available Al alloys for aerospace applications is done using TOPSIS, VIKOR, WASPAS, MOORA, EDAS and CODAS MCDM techniques and to check the consistency in ranking, Spearman rank correlation coefficient was computed. Almost a strong correlation is exhibited and there is no much significant difference in ranking. From the analysis, AA2024 Al alloy with Si (wt. 0.5%), Fe (wt. 0.5%), Cu (wt. 3.8–4.9%), Mn (wt.0.3–0.9%), Mg (wt. 1.2–1.8%), Cr (wt. 0.1%), Zn (wt. 0.25%), Ti (wt. 0.15%), unspecified elements (wt. 0.2%) and remaining major constituent as aluminium with the characteristics of density 2.77 g/cc, elastic modulus of 72.4 GPa, yield strength of 325 MPa, ultimate strength of 470 MPa, % elongation 20 and fatigue endurance limit of 140 MPa; AA7178 Al alloy with Si (wt. 0.4%), Fe (wt. 0.5%), Cu (wt. 1.6–2.4%), Mn (wt.0.3%), Mg (wt. 2.4–3.1%), Cr (wt. 0.18–0.28%), Zn (wt. 6.3–7.3%), Ti (wt. 0.2%), unspecified elements (wt. 0.2%) and remaining major constituent as aluminium with the characteristics of density 2.83 g/cc, elastic modulus of 71.7 GPa, yield strength of 440 MPa, ultimate strength of 572 MPa, % elongation 17 and fatigue endurance limit of 130 MPa can be chosen for aerospace applications with less weight to volume ratio, better tensile and fatigue characteristics. As the base material is selected for metal matrix composite, further it can be processed for reinforcement prospects.

REFERENCES

1. Nturanabo F, Masu L and Kirabira JB (2019) Novel Applications of Aluminium Metal Matrix Composites, Aluminium Alloys and Composites, Kavian Omar Cooke, IntechOpen.
2. Surappa MK (2003) Aluminium matrix composites: Challenges and opportunities. *Sadhana* 28, 319–334.
3. Rajan R, Kah P, Mvola B and Martikainen J (2016) Trends in aluminium alloy development and their joining methods. *Reviews on Advanced Materials Science* 44, 383–397.
4. ASM International Handbook Committee (1992) Properties and selection: nonferrous alloys and special-purpose materials. *ASM International*, 2, 1143–1144.
5. Davis JR (2001) *Alloying: Understanding the Basics, Aluminum and Aluminum Alloys*, ASM International, 351–416.
6. Aboulkhair NT, Simonelli M, Parry L, Ashcroft I, Tuck C and Hague R (2019) 3D printing of aluminium alloys: Additive manufacturing of aluminium alloys using selective laser melting. *Progress in Materials Science* 106, 100578.
7. Dursun T and Soutis C (2014) Recent developments in advanced aircraft aluminium alloys. *Materials & Design* 56, 862–871.
8. Miller WS, Zhuang L, Bottema J, et al. (2000) Recent development in aluminium alloys for the automotive industry. *Materials Science and Engineering: A* 280, 37–49.
9. Starke Jr. EA, Rashed HMMA (2017) Alloys: Aluminum. *Reference Module in Materials Science and Materials Engineering*.
10. Jeya Girubha R and Vinodh S (2012) Application of fuzzy VIKOR and environmental impact analysis for material selection of an automotive component. *Materials and Design* 37, 478–486.

11. Seyed Haeri MN, and Alireza SA (2018) A new multi-criteria decision making approach for sustainable material selection problem: A critical study on rank reversal problem. *Journal of Cleaner Production* 182, 466–484.

12. Noryani M, Sapuan SM and Mastura MT (2018) Multi-criteria decision-making tools for material selection of natural fibre composites: A review. *Journal of Mechanical Engineering and Sciences* 12(1), 3330–3353.

13. Chakrabortya S and Chatterjeeb P (2013) Selection of materials using multi-criteria decision-making methods with minimum data. *Decision Science Letters* 2, 135–148.

14. Mathewa M and Sahua S (2018) Comparison of new multi-criteria decision making methods for material handling equipment selection. *Management Science Letters* 8, 139–150.

15. Bagga P, Joshi A and Hans R (2017) QoS based web service selection and multi-criteria decision making methods. *International Journal of Interactive Multimedia and Artificial Intelligence* 5(4), 113–121.

16. Karande P and Chakraborty S (2012) Application of multi-objective optimization on the basis of ratio analysis (MOORA) method for materials selection. *Materials and Design* 37, 317–324.

17. Vijay Manikrao A, Chakraborty S (2012) Material selection using multi-criteria decision-making methods: A comparative study. *Proc IMechE Part L: J Materials: Design and Applications* 226(4), 266–285.

18. Jahan A, Mustapha F, Ismail MY, et al. (2011) A comprehensive VIKOR method for material selection. *Materials Design* 32, 1215–1221.

19. Mukhametzyanov I and Pamučar D (2018) A sensitivity analysis in MCDM problems: A statistical approach. *Decision Making: Applications in Management and Engineering* 1(2), 51–80.

20. Dehdasht G, Ferwati MS, Zin RM, Abidin NZ (2020) A hybrid approach using entropy and TOPSIS to select key drivers for a successful and sustainable lean construction implementation. *Plosone* 15(2), e0228746.

6 Study of Machinability, Mechanical, and Tribological Properties of Hybrid Al-MMC

M. Vignesh
Amrita School of Engineering, Amrita Vishwa Vidyapeetham

G. Ranjith Kumar
Sri Venkateswara College of Engineering
and Technology (Autonomous)

M. Sathishkumar
Amrita School of Engineering, Amrita Vishwa Vidyapeetham

G. Rajyalakshmi and R. Ramanujam
Vellore Institute of Technology

CONTENTS

DOI: 10.1201/9781003345466-6

6.1   INTRODUCTION

The conventionally available monolithic materials have limitations in attaining properties like high strength, toughness, stiffness, higher density, etc. To overcome this shortcoming and to fulfil the ever-demanding increased material properties for present-day technology, composite has gained considerable interest among the industrial community. Composites are the structural material that possesses two or more constituents to obtain combined and increased properties, which the individual constituent may not attain. The significant difference between an alloy and a composite is that the alloying constituents are soluble in each other and form a newer material with different properties, which may not be the same as that of the constituent element property. In composites, all constituent elements are insoluble and retain their individual properties. The adding constituents in composites are at macroscopic levels, insoluble in each other, and have a recognizable interface between them. The interface decides the property of the newly formed composite material. Typically, a composite material possesses a continuous (matrix phase) and a discontinuous phase (reinforcement phase). The discontinuous phase is usually more robust and acts as a load-bearing element in the composite. In contrast, the continuous phase is softer and acts as a load transfer medium. Composite materials are non-homogeneous and isotropic and whose properties are decided by the property of the constituent elements, its orientation, type, distribution, etc. Based on the need for composite and its application, the type of matrix is varied to obtain different categories of composites like polymer matrix composite (PMC), ceramic matrix composite (CMC), metal matrix composite (MMC), etc.

6.2   METAL MATRIX COMPOSITES

MMCs generally possess metal (magnesium, aluminium, copper, iron, cobalt) as a base matrix and ceramic material (oxides, carbides) as a reinforcing agent. The different composites are formed based on the ceramics' application, requirement, and wettability with the metal matrix. These MMCs are a fascinating choice for the conventionally available materials, which possess improved properties like high wear resistance, high thermal conductivity, higher modulus, good corrosion resistance, better density, etc. MMCs have improved applications in industrial areas like aerospace, defence, automobile, marine, architecture, etc. Compared to conventional MMCs, hybrid metal matrix composites (HMMCs) have gained higher importance. HMMCs are the ones in which two different reinforcing agents are added to the base metal matrix to get new material with improved characteristics [1]. The high-strength composite materials possess enhanced properties when compared to the monolithic alloys. The classification of MMC is given in Figure 6.1. Among the various available metal matrices, aluminium matrices are highly used in the industrial sectors because of its improved properties like a good ratio of strength to weight, toughness, impact resistance, high specific modulus, etc. The melting point of the aluminium material is nominally higher for its place of application and was also found suitable for its processing in the development of composites [2]. Regarding reinforcements, oxides and carbides exhibit excellent corrosion resistance, good thermal resistance, higher strength and modulus, etc. Hence these materials are used as a reinforcing agent to the base aluminium matrix, resulting in the formation of high-strength composite materials.

6.2.1 Advantages of MMCs

- Higher thermal and electrical conductivity
- Higher resistance to impact damage
- Increased compressive strength
- Increased contribution in composite transverse properties
- Significantly higher strength and stiffness
- No product shape and size restriction
- Lower cost

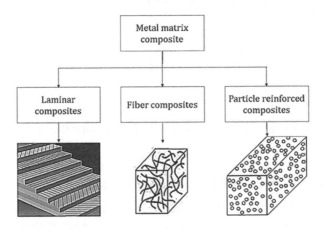

FIGURE 6.1   Classification of MMC.

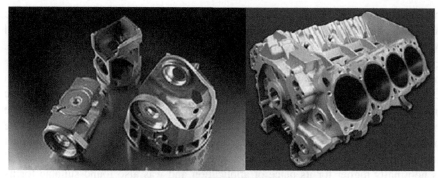

FIGURE 6.2   Applications of MMC.

6.2.2   DISADVANTAGES OF MMCs

- Limited and costlier casting technology
- Complex and inefficient bonding technology
- Higher chances for thermal fatigue
- Conduction fibre reinforcements prone to corrosion
- The fibre-reinforced system has a complex fabrication method

6.2.3   APPLICATIONS

MMCs have larger applications in many industrial sectors; their images are shown in Figure 6.2.

6.3   TYPES OF THE ALUMINIUM BASE MATRIX

Many grades of aluminium are available in the market, from series 1 to series 8. The introduction and development of a newer aluminium series satisfy the drawbacks and shortcomings of the already available older aluminium series. The more recent aluminium series varies from the previous series, with a slight variation in the elemental composition. Every series of aluminium is used in different applications based on its

TABLE 6.1

Different Aluminium Grades and Its Properties

Aluminium Grades	Alloying Elements	Properties	Applications
1xxx	Unalloyed (99% pure aluminium)	Higher corrosion resistance and excellent workability	Good corrosion resistance, workability, electrical and chemical materials
2xxx	Copper	Excellent ratio of strength to weight, fatigue resistance	Aerospace materials
3xxx	Manganese	Higher corrosion resistance and excellent workability	Deep drawing, brazing, welding
4xxx	Silicon	Corrosion resistance, good thermal conductivity, excellent electrical conductivity, higher melting point	Used as welding wires, brazing alloy, and architectural applications
5xxx	Magnesium	Strain-hardenable resists corrosion, readily weldable	Boat hulls, gangplanks, marine equipment
6xxx	Silicon, magnesium	Good formability, weldability, machinability, and poor corrosion resistance	Architectural, marine, and general-purpose applications
7xxx	Zinc	Most potent of all wrought alloys, decrease workability and machinability	Aircraft applications, mobile equipment, highly stressed parts
8xxx	Other elements	Lower densities, higher stiffness, performance at elevated temperatures	Helicopter components, other aerospace applications

property and its practical usefulness at the place of application. The other available aluminium series and its properties are given in Table 6.1.

In considering all these properties of aluminium alloys, many researchers have conducted studies on different series at different periods to check its suitability at the place of application and evaluate the structural changes on the material due to other treatments. Apart from its usage in its present form as alloys, researchers have included some reinforcing agents to it and call them composites. The development of aluminium composites is to obtain a material with higher strength and improved properties at lower cost and lesser weight.

6.4 TYPES OF REINFORCING AGENTS

Reinforcing agents, as the name suggests, are reinforced in the base metal matrix to enhance the material's properties. Also, some of the material's physical properties like thermal conductivity, wear resistance, friction coefficient, etc. are changed. Based on the strength required, the type of reinforcing agents is varied, like fibres (continuous reinforcement), whiskers, and particulates (intermittent reinforcements). Also, the material types of reinforcing agents are different, such as metals, ceramics, etc. Adding ceramic particles as reinforcing agents to the base matrix enhances the strength and melting point of the fabricated composites. Different ceramic reinforcing particles are added to the base matrix based on the property that needs to be improved in the newly forming material. Some of the ceramic reinforcement particles used in composite manufacturing are titanium carbide (TiC), silicon carbide (SiC), titanium diboride (TiB_2), aluminium oxide (Al_2O_3), tungsten carbide (WC), boron carbide (B_4C), etc. Molybdenum disulphate (MoS_2) and graphite are added to increase the lubricity and wettability of the composite [3]. Boron carbide and silicon carbide are cost economical and possess improved properties like good wear resistance and specific stiffness. The titanium carbide reinforcement has excellent chemical inertness and a higher melting point and exhibits higher strength and hardness at elevated temperatures, which are the desirable properties during composite fabrication.

6.4.1 DISCONTINUOUS REINFORCEMENTS

Usually, in composites, as the length of reinforcement increases, the mechanical properties like strength and stiffness of the composites increase. The whisker particles possess a higher aspect ratio when compared to the particulate reinforcements. Hence the strength of the whisker is found to be lesser than particulate reinforcements [4]. During the processing of composites, these whiskers tend to break into smaller lengths, thus making its strength ineffective. Also, the whiskers tend to orient in different directions, exhibiting different properties. At the same time, the particulate reinforcements show uniform property in all directions.

6.4.2 CONTINUOUS REINFORCEMENTS

Fibres come under continuous reinforcements, which contain carbon and ceramic as the continuous reinforcements. Commonly used carbon fibre is graphite and polyacrylonitrile (PAN) precursor. The ceramics used as continuous reinforcements are B_4C, TiB_2, SiC, SiO_2, Al_2O_3, zirconia, mullite, etc. [5]. All the fibres mentioned above are brittle and flaw sensitive; they also function based on the size effect of the fibres. Hence, as the length of the fibre increases, the strength decreases because of the reduced fibre size used for composite fabrication. One more type of continuous reinforcement is wire reinforcement, which is rarely used for composite fabrication. These wire reinforcements are highly utilized in specific niche applications based on its suitability for the said application. Some wires used as reinforcements are tungsten, stainless steel, titanium, beryllium, molybdenum, etc. For instance, tungsten wire-reinforced composites show high-temperature creep resistance, which found its applications in fighter jet engines and other aerospace components.

The conventionally available MMC consists of single reinforcement, which could be in any form. As there is a need for an increase in strength of the produced composites, adding more than one reinforcement to the base matrix is done, and it is called hybrid metal matrix composites (HMMCs). In the present industrial sectors,

HMMCs have attained higher importance because of their enhanced properties at lower cost, whose strength is the same as that of the conventionally available high-strength, high-cost material. The improved properties like low density, high temperature, high fatigue strength, wear resistance, hardness, stiffness, and lightweight properties have attracted many industrial sectors to use the HMMCs as the choice of material for extreme applications. Because of the light property, the use of HMMCs in the aircraft industry has gained tremendous interest, reducing the use of increasing fuel consumption in the aircraft.

Among the available HMMCs, aluminium-based HMMCs [6] are highly used in the industrial sectors because of their compatibility with all the reinforcing agents and their lightweight material capability. Hence, many researchers have focussed on studying machinability, mechanical, and tribological properties of hybrid aluminium MMCs. Therefore a detailed study of the selected material would gain increased interest among the research community.

6.5 HYBRID METAL MATRIX COMPOSITES

To have a thorough knowledge of the properties and advantages of the aluminium-based HMMCs, many researchers have selected various grades of aluminium as a matrix and reinforced them with different kinds of reinforcing agents. The reinforcing agents added varied at different percentages to check their strength and suitability for the application. As the application of aluminium-reinforced hybrid composites increases, reinforcements are also varied from micro-level to nano-level. The change or variation in the size of the reinforcements is to check the strength enhancement of the HMMCs.

Parakh Agarwal et al. [7] used aluminium (Al) 7075 as the base matrix, reinforced with hexagonal boron nitride (h-BN) and graphene nano-particles. The composites were prepared using the squeeze cast process, and the added reinforced particles were mixed in the ball milling process before adding them to the base matrix. The prepared composites were studied for its mechanical property improvement over unreinforced Al 7075 alloy and found increased properties on micro-hardness and ultimate tensile strength by 10.93% and 31.25%, respectively. Also, the fabricated composites are studied for its machinability properties like forces generated, machined surface quality, wear observed on the tool surface and the type of chip formed during machining. The machining experiments are carried out in the dry and minimal quantity lubrication (MQL) environment and found that MQL environment machining fetches desired results than dry machining.

A similar machinability study on a composite of Al 7075 reinforced with nano-alumina and h-BN particles is reported by Kannan et al. [8]. The sample fabrication is carried out through a squeeze cast process possessing a pressure of 150 MPa. The influence of machining environment (dry and MQL) on the cutting force generated, quality of machined surface, and the amount of wear generated on the cutting tool during machining are studied. Also, the reinforced composites and unreinforced alloy are compared for the selected responses. The nano-composites exhibit better machinability in the MQL environment in terms of reduced cutting force, surface roughness, and tool wear.

Thirumalai Kumaran et al. [9] used Al 6351 aluminium alloy as the base matrix with silicon carbide and boron carbide as the reinforcing agents. The addition of reinforcements makes the composite material the hard-to-cut material, which attracts the researchers to analyse more on the machinability aspects of the fabricated composites. Some of the machinability aspects selected by the said researchers are the material removal rate, surface roughness generated during machining, and the amount of power consumed. The results proved that variation in boron carbide reinforcement percentages resulted in a change in surface roughness and power consumption with 7.87% and 6.36%, respectively. At the same time, the increase in the material removal rate deteriorates the machined surface quality and the amount of power consumed during machining.

Engin Nas and Hasan Gokkaya [10] fabricated hybrid composites with B_4C and graphite particles added at a micro-level and Alumix 13 as the base matrix. These graphite powders with nickel coating were varied at 0, 3, 5, and 7 weight percentages, and B_4C was added at 8 constant weight percentages. The various machinability studies are performed on the fabricated composites with varying feed rates, cutting speed, and a constant depth of cut. From the results, the different weight percentages of composites contribute more to deciding the machined component quality, followed by cutting speed and feed rate. The most critical and challenging output of MMC machining is the built-up edge (BUE) formation on the tool. It is formed due to the addition of aluminium material on the tool surface, resulting in poor machined surface quality and causing rapid tool wear. The BUE formation is highly contributed by the type of material used, cutting speed, feed rate, and the depth of cut.

While the authors use graphene particles, hexagonal boron nitride, alumina, silicon carbide, graphite, and boron carbide, as reinforcing agents, Shoba et al. [11] use rice husk ash (RHA) and SiC as the reinforcements at a micro-level to the base Al 356 alloy. The composites are analysed for their machinability properties like surface quality, forces generated, and wear observed on the cutting tool. The cause of BUE formation on the tool and its effect on the work material is also studied, and it was found that a higher cutting speed produces a good surface finish without BUE. For the wear mechanism is concerned, abrasion type wear is prevalent during machining the fabricated composites. All the observed experimental results are validated through analytical equations.

In addition to the machinability property, the need for the use of HMMCs is their enhanced mechanical and physical properties. Hence the study on the said properties has attracted the increased attention of many researchers. Rajmohan et al. [12] used SiC and mica particles as the reinforcing agents to the base Al 356 alloy. The mica reinforced composites show good wear resistance and increased density compared to SiC reinforced composites. The load parameter in wear analysis is the most critical factor in wear loss determination, followed by the mass fraction of the mica.

Also, the varying volume fractions of rice husk ash at 2, 4, 6, 8 wt% with SiC at constant weight percentage are added to Al 356 alloy, and properties like porosity, density, and hardness are studied. It was observed that porosity and hardness increase with a reduced density as the particle content increases. Also, the yield strength and ultimate tensile strength increase as the reinforcement increases with a reduced elongation percentage. Further, the hardness of the composites is increased by performing the aging process [13].

Arun Premnath et al. [14] used a constant weight percentage (5%) of graphite with varying Al_2O_3 percentages (5%, 10%, 15%) to the base Al 6061 alloy, and its mechanical and tribological properties are evaluated. As the Al_2O_3 percentage increases, the density and hardness increase. For the wear analysis is concerned, the load is the most contributing factor in deciding the specific wear rate, followed by speed and reinforcement percentages. On examination of the worn-out surface of the composite pin, the wear mechanism type observed is mainly found to be abrasive, followed by oxide wear.

Mohan Kumar et al. [15] used Al 430 reinforced with MgO and SiC particles to evaluate wear and mechanical behaviour. The results showed that increased reinforcement weight percentage improved tensile property and hardness. As the sliding distance and load increase, the composite wear rate also increases. It is found to be the opposite of the coefficient of friction.

6.6 PROCESSING TECHNIQUES

MMCs can be done using three techniques: solid-state processing, liquid-state processing, and in-situ processing. The further classification of processing techniques is given as a flowchart in Figure 6.3.

6.6.1 SOLID-STATE PROCESSING

The solid-state processing is carried out to produce the composites without melting the base metal matrix. It is classified into three variants: powder blending and consolidation, diffusion bonding, and physical vapour deposition.

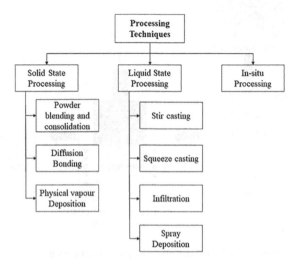

FIGURE 6.3 Processing techniques of MMCs.

i. Powder blending and consolidation

The metal powder is mixed thoroughly with whiskers/particles/short fibres in dry or liquid form. After mixing, it is subjected to cold compaction, degassing, canning, and high-temperature consolidation. The oxide particles formed in the mixture are due to the powder type or the processing technique involved. The formed oxides are very useful in the dispersion strengthening of the MMCs. This technique is highly used in fabricating aluminium and magnesium matrix composites [16]. The steps involved in powder blending and consolidation are shown in Figure 6.4.

ii. Diffusion bonding

This technique is highly used in the fibre-reinforced MMC, with aluminium and magnesium as the base matrix. As the name suggests, it is the inter-diffusion bonding between the atoms at the metallic surfaces and the fibre reinforcements under pressure. The type of fibre reinforcements used could be continuous or discontinuous.

iii. Physical vapour deposition

The reinforcements to be added to the base metal matrix are passed inside the high partial pressure region, where the vapour is passed. The condensation occurs on the reinforcements at a rate of 5–10 micro-metres per

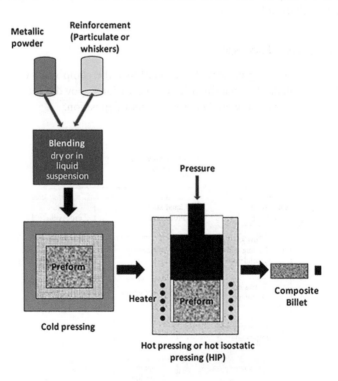

FIGURE 6.4 Steps involved in powder blending and consolidation.

minute by hot pressing and hot isostatic pressing (HIP). The consolidated reinforcements are further used for the composites fabrication.

6.6.2 LIQUID-STATE PROCESSING

The base metal matrix is melted in this process, and the reinforcing agents are added. Further, it is allowed to solidify for the formation of required composites with lesser porosity. It is classified into four variants: squeeze casting, stir casting, infiltration process and spray deposition.

i. Stir casting

In this process, the particulates are added to the liquid metal, and the added mixture is allowed to solidify. The added reinforcements are pre-heated to a specific temperature, preventing the particulates from getting clustered. The drawback of the stir casting process is that the uniform distribution of reinforcements in the matrix is absent. Hence, a maximum of 30% reinforcement addition to the base matrix is recommended for effective composite fabrication. The added reinforcement particles could be of micro- or nano-size in scale. Also, the stirring of molten material plays a vital role in deciding the uniformity of the added reinforcements to the base matrix for the composite fabrication.

ii. Squeeze casting

In this process, the composite mixture is subjected to pressure within the die after introducing the molten metal and reinforcement mixture into the open die. The higher temperature in the composite is transferred to die under pressure, allowing it to solidify. Fabricating composites under pressure results in producing components with fine grain structure and less porosity [17]. The squeeze casting process steps are given in Figure 6.5.

iii. Infiltration process

This composite fabrication method infiltrates the molten metal into the selected reinforcement (fibres/whiskers). Based on the porosity level, the

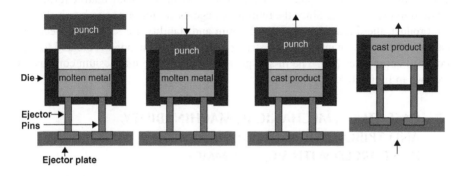

FIGURE 6.5   Squeeze casting process steps [17].

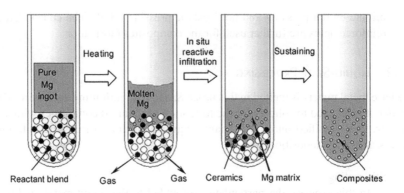

FIGURE 6.6   In-situ process image of the magnesium composites [18].

volume fraction chosen for the composite fabrication is decided. According to the porosity level, the molten metal infiltration into the reinforcing agent is conducted. The silica and metal mixtures are added as binders to retain the shape and integrity of the fabricated composites.

iv. Spray Deposition

In this process, the composites are fabricated layer-by-layer by deposition particles/whiskers/short fibres with 5%–10% porosity on the metal surface, using the spraying technique. The available porosity is reduced later by further processing during fabrication. The matrix material is sprayed onto the fibre surface for continuous or long fibres. The fibre spacing and layer decide the fibre volume fraction in the fabricated composites.

6.6.3   IN-SITU PROCESSING

In the in-situ process, the reinforcement particles are generated through a chemical reaction. This prevents the risk of contamination in the metal matrix due to reinforcements. The precipitation of liquid or solid in the metal matrix forms the required reinforcements. Since the reinforcing agents are formed as a chemical reaction product, the thermodynamic equilibrium and stability are formed between the matrix and reinforcements. Hence a strong matrix dispersion bond is created for the produced composites [18]. The in-situ process image of the magnesium composites is shown in Figure 6.6.

6.7   CASE STUDY: MECHANICAL, MACHINABILITY, AND TRIBOLOGICAL STUDY OF AL 6061 REINFORCED WITH SiC, TiC HMMCs

The properties and performance of the HMMCs are studied with Al 6061 material as the base matrix. It is reinforced with silicon carbide and titanium carbide as the reinforcing agents. The mechanical, machinability, and tribology properties of the fabricated composites are analysed. The composites fabrication is performed using one of the liquid processing techniques named the stir casting route.

6.7.1 FABRICATION BY STIR CASTING ROUTE

The preheated reinforcement particles, SiC and TiC of 20–40 microns in size, are reinforced to the molten melt in the furnace of 800°C, and it is stirred continuously. The magnesium billets of 1 wt% are introduced to the composite mixture to increase the wettability between the molten metal and the ceramic reinforcing agents. Hexa chloro-ethane is added in the powder form to the molten mixture, acting as the degasser in the molten mixture. Also, the composite mixture is ensured to remove captured gases, and the void formation is avoided completely. Three different percentages of composite mixtures are produced, namely, Al 6061 + 5 wt % of SiC + 5 wt% of TiC, Al 6061 + 4 wt% of SiC + 2 wt% TiC, Al 6061 + 2 wt% of SiC + 4 wt% TiC, respectively. After stirring, the molten mixture is poured into the open die of 50 mm in diameter and 300 mm in length. Upon pouring, the composite rod is subjected to squeeze casting for the grain refinement and the removal of trapped gases and the voids in it. The composite is allowed to solidify under the applied pressure, and three different rods of varying percentages and required dimensions are obtained. The experimental setup used for composite fabrication is given in Figure 6.7.

FIGURE 6.7 Experimental setup used for composite fabrication.

6.7.2   MICROSTRUCTURAL EXAMINATION OF THE FABRICATED COMPOSITES

The microstructural examination is the most important and primary characterization study for the MMCs immediately after fabrication. The microstructural analysis needs to check the base matrix's uniform reinforcement distribution and ensure the metal matrix's cluster-free reinforcements. The grain boundaries between the matrix and the reinforcements decide the composites' strength. The microstructure of the fabricated composite was studied by following a standard metallographic procedure. The sliced composite sample is hot pressed using the Bakelite powder for holding and effectively polishing the mounted samples. Polishing needs to be carried out on the increasing grit size of the emery sheets starting from 200 to 2500 grid size. After emery sheet polishing, the disk polishing needs to be performed using velvet cloth and alumina powder to attain glass finish samples. The etchant for the selected aluminium grade is Keller's reagent (3 ml of HCl + 5 ml of HNO$_3$ + 190 ml of H$_2$O) [19], which is applied to the polished surface through a swab process using cotton. It is immediately washed with water and dried to avoid over-etching by the applied etchant. The polished and etched samples are checked for their microstructure using a high-resolution optical microscope, and the image is captured using a machine vision system. The microstructure of the fabricated composites is given in Figure 6.8.

The microstructure of the fabricated composite clearly shows the presence of TiC and SiC in the base matrix. The agglomeration of reinforcements is observed in the microstructure image. The influence of added reinforcements on various properties will be discussed in the forthcoming sections.

6.7.3   MECHANICAL PROPERTY STUDY

Upon reinforcement additions to the aluminium matrix, the entire property of the base alloy gets changed (might be higher or lower) according to the property of the added reinforcements. Also, the newly obtained property of the developed composites needs to be studied and analysed for recommending it for the application. Usually, a material being used at the place of application undergoes mechanical stress

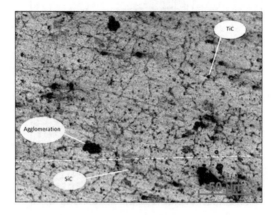

FIGURE 6.8   Microstructure of the fabricated composites.

and loading. Hence, the produced material should be without extreme temperature, pressure, stress, fatigue, etc. [20]. The composites for a specific application should always consider the said factors and be produced. Even upon fabrication, it should be subjected to rigorous mechanical testing, so repeatability of the obtained properties should be delivered. In the present case study, some of the fabricated composites' mechanical properties (impact test, micro-hardness) are studied and analysed.

6.7.4 IMPACT TEST

The impact test gives the material's toughness, impact strength, and fracture resistance by measuring the amount of energy absorbed before fracture. This also checks whether the material is ductile or brittle. This test is mainly performed to analyse the shock loads that material could withstand before it undergoes deformation, fracture, or rupture. The most commonly used impact tests are the Izod test, Charpy test, and tensile impact test. To study the impact strength of the Al HMMCs, the Izod impact test was performed with a sample of $55 \times 10 \times 10$ mm, following the ASTM E23 standard. The images of the samples before and after testing is given in Figure 6.9.

FIGURE 6.9 Samples before (a) and after testing (b) Al 6061 + SiC (4%) + TiC (2%).

TABLE 6.2
Impact Strength Values of Different Composites

Sl. No.	Samples	Notch Impact Strength (J/mm^2)
1	Al 6061 + SiC (5 wt%) + TiC (5 wt%)	0.653
2	Al 6061 + SiC (4 wt%) + TiC (2 wt%)	0.393
3	Al 6061 + SiC (2 wt%) + TiC (4 wt%)	0.565

The impact energy of all the fabricated composite materials is given in Table 6.2. From the results, it is clear that samples with higher reinforcement percentage (SiC (5 wt%) + TiC (5 wt%)) show increased impact strength (0.653 J/mm^2) when compared to the other two specimens. Based on the fracture obtained in the samples after impact testing, the failure mode is found to be brittle. The other two specimens show a lower impact strength of 0.393 J/mm^2 and 0.565 J/mm^2 for SiC (4 wt%) + TiC (2 wt%) and SiC (2 wt%) + TiC (4 wt%) reinforced composites, respectively. Normally, failure at the work material surface starts from the crack zone and propagates further till fracture. Other factors contributing to fracture are the blow holes and the porosity formed during composite fabrication. This shows that the increase in reinforcement percentage to the base aluminium matrix exhibits increased toughness. Also, the material used for the application will withstand the said load during operation.

6.7.5 Micro-Hardness Study

Hardness is the property of the material which shows resistance to abrasion and scratch on the material surface when it is subjected to extreme mechanical operations at the zone of application. Usually, ceramic materials show excellent hardness individually, which upon addition to aluminium alloy matrix possess higher hardness than the self. This enhances the self-hardness and improves the property of the aluminium alloy to be used in extreme applications. The selected reinforcements' compatibility with the particular base alloy should be checked before its addition. Improper selection and addition of the reinforcing agents might reduce the base material property rather than enhance it. Also, the hardness present in the material should be uniform throughout the material, from surface to bulk. The difference in hardness in the same material would result in premature failure of the component instead of its rated working life period. The addition of reinforcement to the base alloy should be done in a proper and required percentage because the increased addition of reinforcements not only enhances the material hardness but also increases the brittleness of the component. This would prevent the material from being utilized at the higher and sudden impact zone of applications.

In the selected case study, SiC and TiC are added as the reinforcement particulates to the base Al 6061 alloy. The percentage of the reinforcements are varied as 5 wt % of SiC + 5 wt% of TiC, 4 wt% of SiC + 2 wt% of TiC, 2 wt% of SiC + 4 wt% of TiC, respectively. The composite materials are studied for their hardness property by measuring its micro-hardness on Vicker's scale. The measurement done on the fabricated composites should have a uniform hardness throughout the material. This confirms the uniform mixture and distribution of reinforcements in the base alloy. The added reinforcements are on a micro-scale; hence, the hardness measurement at the macrolevel would damage the work material. This encourages using micro-level hardness measurement on the fabricated work material. The micro-hardness measurement instrument used is Make: Matsuzawa with a diamond as the indentor material. The selected load for testing is 200 gf, which is applied for 10 seconds at a distance of 500 µm. The sample subjected to micro-hardness testing is neatly polished to a mirrorlike finish and will be neatly mounted on a good base. The experimental setup used and the mounted sample are shown in Figure 6.10.

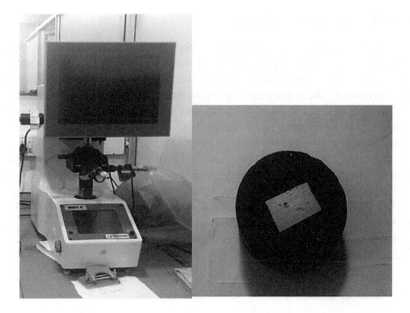

FIGURE 6.10   Micro-hardness measurement instrument with the mounted sample.

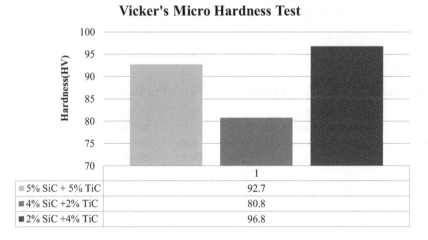

FIGURE 6.11   Micro-hardness values obtained for the fabricated composites.

The micro-hardness value obtained is given in Figure 6.11. The graphs show that Al 6061 reinforced with 2% SiC and 4% TiC exhibits excellent and higher (96.8 HV) micro-hardness than the other two specimens. The least micro-hardness of 80.8 HV is obtained for Al 6061 with 4% SiC and 2% TiC reinforcements. Based on the results obtained, the increase in reinforcement percentage of TiC with reduced SiC reinforcement percentage gives increased micro-hardness because of the higher load-bearing capacity of the added reinforcements.

6.8 MACHINABILITY STUDY

The fabricated composite materials possess increased hardness, strength, toughness, etc. Because of these improved properties, it would be challenging to machinability. In the prepared three compositions of composite materials, some machinability studies are performed, like cutting force analysis, surface roughness measurement, and tool wear studies. Cutting speed, feed rate, and material type are chosen as varying parameters at three levels to carry out the machinability studies. The experimental design selected for the experiments is given in Table 6.3.

In the table, the term A represents Al+TiC(5%)+SiC(5%); B represents Al+SiC(4%)+TiC(2%); and C represents Al+SiC(2%)+TiC(4%). Trials are conducted based on the input conditions to evaluate the above machinability properties. Also, based on it, the components utilized at the application area should be fabricated with increased accuracy, precision and surface finish. The cutting tool used was coated carbide with a coating order of TiN-Al$_2$O$_3$-TiCN-TiN, with cemented carbide as the base substrate.

6.8.1 SURFACE ROUGHNESS

The first and foremost requirement of the customer in selecting a component is its aesthetics and appearance. So, the surface finish of the product that is fabricated should be given higher importance than its dimensional accuracy. The component with good surface quality attracts more users and enhances the selected part's

TABLE 6.3
Experimental Design for Conduct of Machinability Study

Run Order	Cutting Speed (m/min)	Feed Rate (mm/rev)	Sample Type	Run Order	Cutting Speed (m/min)	Feed Rate (mm/rev)	Sample Type
1	60	0.1	A	15	120	0.15	B
2	90	0.1	A	16	60	0.2	B
3	120	0.1	A	17	90	0.2	B
4	60	0.15	A	18	120	0.2	B
5	90	0.15	A	19	60	0.1	C
6	120	0.15	A	20	90	0.1	C
7	60	0.2	A	21	120	0.1	C
8	90	0.2	A	22	60	0.15	C
9	120	0.2	A	23	90	0.15	C
10	60	0.1	B	24	120	0.15	C
11	90	0.1	B	25	60	0.2	C
12	120	0.1	B	26	90	0.2	C
13	60	0.15	B	27	120	0.2	C
14	90	0.15	B				

working life. The smooth finish products will have a crack and defect-free surfaces, resulting in higher fatigue life and load-bearing capacity and increasing the scratch and abrasion resistance of the component in high-strength jobs [21]. Many factors decide the surface quality of the component, such as the proper selection of tools, the vibration-free base of the machine during machining, operator attentiveness, periodical maintenance of the machine, exact selection of machining conditions, etc. The surface quality is calculated by measuring the average surface roughness (Ra), the average of all peaks and valleys available in the roughness profile.

The machinability study is performed on all three fabricated composite materials, and its respective average surface roughness is measured and plotted. The surface roughness measurement is conducted on the Mahr Surf surface roughness profiler with a cut-off distance of 5.6 mm. The measurement is performed thrice for all the specimens, and experimental runs and the average values are plotted.

The average surface roughness with varying cutting speed and constant feed rate are measured and plotted in Figure 6.12. Overall, based on the plot, it is clear that an increase in cutting speed and feed rate decreases the surface roughness. At 0.15 mm/rev of feed rate, the surface roughness increases as the cutting speed increases for two specimens. The reason for variation in a trend of surface pattern is due to reinforcement percentages and the non-uniform distribution of the particles.

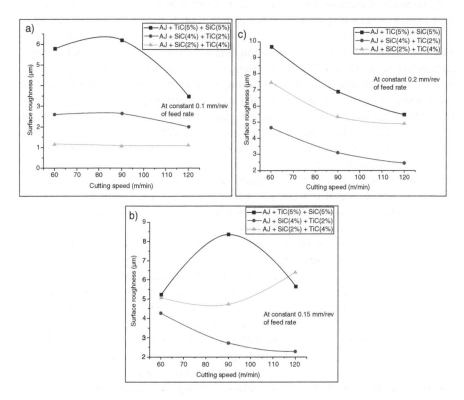

FIGURE 6.12   Average surface roughness at varying cutting speeds and constant feed rate.

These particles are subjected to pullout during machining and cause a rough surface. Another criterion for increased surface roughness is porosity formed during sample fabrication. Among the three composites, the sample with higher reinforcing agents Al + TiC (5%) + SiC (5%) exhibits higher surface roughness irrespective of the feed rate chosen. Out of the three specimens, Al + TiC (2%) + SiC (4%) specimen shows the lowest surface of 1.092 μm at 90 m/min of cutting speed and 0.1 mm/rev feed rate.

6.8.2 Cutting Force

The force exerted by the cutting tool over the workpiece to remove work material from its surface is called cutting force. For effective machining, the cutting tool used for material removal should possess higher hardness than the work material. The cutting force is a significant factor that must be considered by all the machining industries and researchers working in the machining sector. In conventional machining processes, the tool would contact the work material for material removal. During such a process, more stress is induced in the work material. If the generated stresses are compressive, the process is highly acceptable. Also, the said forces create a lot of tool wear on the cutting tool used. Cutting or machining conditions are to be selected to generate a lesser amount of force or heat in the tool and work material [22]. Because of increased force generation, more heat would be created, which in turn causes increased tool wear on the cutting tool. If the worn-out tool is further used in the machining operation, it causes poor surface on the work material surface and results in increased surface roughness. Hence, a proper selection of input parameters is highly needed for obtaining desired and required output responses. Three types of forces are generated during the turning process: thrust force, feed force, and cutting force, respectively, and are generated perpendicular to each other. Out of the three forces, the cutting force is given higher importance which acts in the Z-direction. Hence, the cutting forces are analysed in detail in the present case study.

During machining of aluminium-based composites, the force generated would vary according to the type of reinforcement added. Ceramic reinforcements are stiffer, which upon addition to aluminium alloy would result in increased hardness. Hence, the machining of it would also be a challenging one. The machining operations are performed in the selected aluminium-based composites with three compositions of reinforcing agents to analyse the forces generated during machining. Twenty-seven experiments were conducted with cutting speed, and feed rate, varying at three levels and the results obtained are plotted and given in Figure 6.13.

In the given plot, it is clear that as the feed rate increases, the cutting force generated during machining also increases. As for composite material, the composite with Al as base material and TiC and SiC each varied at 5%, resulting in the highest cutting force, irrespective of the feed rate. The least cutting force of 68.3 N is generated for the composite, Al, with 2% SiC and 4% TiC as reinforcement, at 0.1 mm/rev of feed rate and 90 m/min of cutting speed. The highest cutting force of 281.5 N is generated for Al with 5% SiC and 5% TiC as reinforcements, at a higher cutting speed (120 m/min) and feed rate (0.2 mm/rev) among the selected machining conditions. Though both the reinforcements are ceramic, SiC exhibits increased property

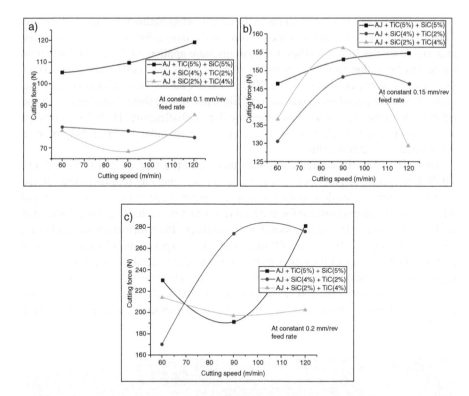

FIGURE 6.13 Cutting force at varying cutting speeds and constant feed rate.

when compared to TiC particles. So, the composites with a lesser percentage of SiC reinforcement (2%) show lesser cutting force when compared to other composites.

6.8.3 TOOL WEAR

The maximum cost is spent on tooling used in machining required components in the manufacturing and machining industries. Almost 50% of the component fabrication cost is spent on tooling itself, followed by material and labour. A proper selection of tools for the specific work material is to be done for effective machining of the selected work material. As the use of high-strength materials increases, new tools have entered the machining market. Some new high-strength cutting tools are polycrystalline diamond (PCD), cubic boron nitride (CBN), ceramics, coated tools, etc. That particular tool is selected according to the type of material to be machined using a selected machining process. But these said tools are higher in cost, which would further result in higher machining cost of the material. To overcome this, multi-layer coatings on carbide base tools are introduced [23]. Based on the improved performance of the coatings, many industrialists have started using the coated tools and made the processes economical.

Many coatings are available, namely, titanium nitride, titanium carbide, titanium carbo-nitride, aluminium oxide, titanium aluminium nitride, etc. Each coating layer has properties like increased toughness, hardness, temperature bearing capacity, etc. The selection and application of the coating on the base substrate vary based on the material to be machined. Earlier, the coatings were applied on the carbide substrate alone, but later, due to the introduction of newer and high-performance materials, the coatings were performed on the PCD and CBN-based tools. This is because the properties of advanced materials are very high; machining them using bare PCD and CBN tools makes the process uneconomical and challenging. High hard coatings are performed on the selected tool material surface to make the machining process economical to overcome this.

To machine the fabricated composite materials, the tool used was TiN-Al$_2$O$_3$-TiCN-TiN coated carbide tool. The presence of SiC and TiC reinforcements in the composites causes abrasion on the tool material surface due to the contact of the tool with the work material surface for material removal. On machining, it was found that BUE was formed on the cutting tool in a wedge shape. BUE formation was higher for lower machining conditions like 60 m/min of cutting speed and 0.1 mm/rev of feed rate. The contact time of the tool with that of the work material surface is found to be high at lower machining conditions. This also causes increased temperature generation at the machining zone and causes coating loss. Also, the work material adhesion on the tool surface is higher due to the increased contact time. At higher machining

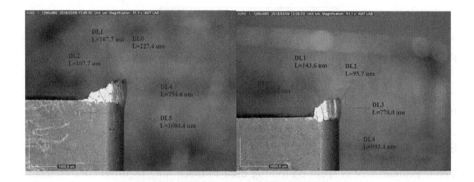

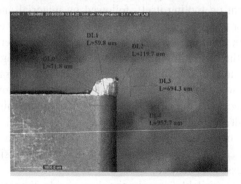

FIGURE 6.14 Tool images after machining of three composites.

conditions, the built-up edge formation is lower; at some machining conditions, it is negligible. The images of the tool with built-up edge formation on machining different composite materials are given in Figure 6.14.

6.9   TRIBOLOGY STUDY – WEAR ANALYSIS

Generally, aluminium is used in many lower application areas because of its lightweight capacity and durability. Use of it at higher wear and tear applications is wholly neglected because of its insufficient wear resistance capacity. To enhance its utility in a vast area of applications, the addition of reinforcements to the base alloy is highly recommended and is in use in the present-day scenario. Typically the reinforcements used are ceramic with higher hardness than the base aluminium alloy. Hence, adding it increases the composite material's hardness, improving the fabricated specimens' wear resistance. Coefficient friction is another response that needs to be concentrated while determining the wear rate of the specimen. It is the relationship between the two objects and the normal drawn for the two objects. The customarily used equipment for the measurement of the wear rate of the specimens is a pin on the disc apparatus. The test sample is taken as a pin, and the application area is taken as a disc. The factors that need to be considered for wear rate measurement are the sliding speed, sliding distance, and load [24]. The coefficient of friction and wear rate are calculated by varying the input parameters. As the sliding speed increases, the material's surface becomes smooth and its temperature increases. This causes a hardening effect on the work material surface and reduces the wear rate. The specimen with a rough surface possesses a higher wear rate and causes easy damage and deterioration of the material used at the application.

With all the input conditions, the pin on disc wear measurement is conducted for the fabricated composite materials with three different compositions. The sliding speed and load are kept at 2 m/s and 6 kg, respectively. The sliding distance alone is varied at three levels of 2000, 4000, and 6000 m. Experimentation showed that the composite with a higher SiC percentage possesses higher wear resistance when compared to TiC [25]. Hence the sample with Al + 5% SiC + 5% TiC exhibits improved wear resistance because of the increased hardness of the ceramic materials used in the said composites. The abnormalities present in the wear testing results are due to defects obtained in the work material during sample fabrication [26]. The commonly obtained defects which affect the wear resistance are the porosity, clustering of reinforcements, etc. Hence, it could be inferred that SiC particles have improved

TABLE 6.4
Coefficient of Friction Values

Sample	Coefficient of Friction
Al + 5%TiC + 5%SiC	0.334
Al + 2%TiC + 4%SiC	0.347
Al + 4%TiC + 2%SiC	0.363

FIGURE 6.15   Specimens before and after wear testing.

significantly in determining the wear resistance of the work material [27]. The coefficient of friction is calculated and tabulated in Table 6.4. The images of the sample before and after wear testing are given in Figure 6.15.

The figures clearly show the samples before and post-wear testing, with neat and clear wear tracks on the specimen. Also, the presence of porosity and blow holes in the fabricated specimens is visible, which reduces wear resistance on the HMMCs.

6.10   CONCLUSION

The present book chapter deals with the complete details on different types of aluminium alloys used in today's market for various applications like structural, automobile, mining, aircraft, etc. Different available base matrix reinforcements to enhance the fabricated composites' strength are discussed elaborately. The complete details of varying fabrication methods available for generating HMMCs are given sumptuously. After providing the basic information on composites and their constituents, a small case study was taken to discuss the results of machinability and mechanical and tribological properties of Al 6061 alloy reinforced with SiC and TiC particulates. The results proved that the developed hybrid composites (Al + SiC + TiC) possess enhanced strength and property on the said responses. The results obtained during the tests are discussed in detail with suitable illustrations.

REFERENCES

1. Palanikumar K, Muniaraj A (2014), 'Experimental investigation and analysis of thrust force in drilling cast hybrid metal matrix (Al–15%SiC–4%graphite) composites', *Measurement*, 53, 240–250.
2. Necat Altınkok (2015), 'Investigation of mechanical and machinability properties of Al2O3/SiCp reinforced Al-based composite fabricated by stir cast technique', *J Porous Mater*, 22(6), 1643–1654.
3. Venkatesan K, Ramanujam R, Joel J, Jeyapandiarajan P, Vignesh M, Darsh J T, Venkata Krishna R, (2014), 'Study of cutting force and surface roughness in machining of Al alloy hybrid composite and optimized using response surface methodology', *Procedia Eng*, 97, 677–686.
4. Sahoo A K, Pradhan S, Rout A K, (2013), 'Development and machinability assessment in turning Al/SiCp-metal matrix composite with multi-layer coated carbide insert using Taguchi and statistical techniques', *Arch Civ Mech Eng*, 13, 27–35.
5. Latha Shankar B, Anil K C, Prasann J Karabasappagol J, (2016), 'A study on effect of graphite particles on tensile, hardness and machinability of aluminium 8011 matrix material', *IOP Conf. Series: Mater Sci Eng*, 149, 012060.
6. Chamarthi P, Nagadolla R, (2019), 'Grey fuzzy optimization of CNC turning parameters on AA6082/Sic/Gr Hybrid MMC', *Mater Today: Proceed* 18, 3683–3692.
7. Agarwal P, Kishore A, Kumar V, Sourabh Kumar S and Thomas B, (2019), 'Fabrication and machinability analysis of squeeze cast Al 7075/h-BN/graphene hybrid nanocomposite', *Eng Res Express*, 1, 015004.
8. Kannan C, Ramanujam R, Balan A S S, (2017), 'Machinability studies on Al 7075/BN/Al2O3 squeeze cast hybrid nanocomposite under different machining environments', *Mater Manuf Process*, 33(5), 587–595.
9. Thirumalai Kumaran S, Uthayakumar M, Slota A, Aravindan S and Zajac J, (2015), 'Machining behavior of AA6351-SiC-B4C hybrid composites fabricated by stir casting method', *Particul Sci Technol*, 34(5), 586–592.
10. Nas E, Hasan Go K, (2017), 'Experimental and statistical study on machinability of the composite materials with metal matrix Al/B4C/graphite', *Metall Mater Trans A*, 48, 5059–5067.
11. Chintada S, Siva Prasad D, Raju P, (2019), 'Investigations on the machinability of Al/SiC/RHA hybrid metal matrix composites', *Silicon*, 11, 2907–2918.
12. Rajmohan T, Palanikumar K, Ranganathan S, (2013), 'Evaluation of mechanical and wear properties of hybrid aluminium matrix composites', *Trans Nonferrous Met Soc China*, 23, 2509–2517.
13. Siva Prasad D, Shoba C, Ramanaiah N, (2014), 'Investigations on mechanical properties of aluminium hybrid composites', *J Mater Res Technol*, 3(1), 79–85.
14. Arun Premnath A, Alwarsamy T, Rajmohan T, Prabhu R, (2014), 'The influence of alumina on mechanical and tribological characteristics of graphite particle reinforced hybrid Al-MMC', *J Mech Sci Technol*, 28 (11), 4737–4744.
15. Mohan Kumar S, Pramod R, Govindaraju H K, (2017), 'Evaluation of mechanical and wear properties of aluminium AA430 reinforced with SiC and MgO', *Mater Today: Proceed*, 4, 509–518.
16. Contreras Cuevas A, Bedolla Becerril E, Martínez M S, Lemus Ruiz J, (2018), '*Fabrication Processes for Metal Matrix Composites', Metal Matrix Composites.* Springer, Cham.
17. Ghomashchi M R, Vikhrov A, (2000), 'Squeeze casting: An overview', *J Mater Process Tech*, 101, 1–9.

18. Dong Q, Chen L Q, Zhao M J, Bi J, (2004), 'Synthesis of TiCp reinforced magnesium matrix composites by in situ reactive infiltration process', *Mat Lett*, 58(6), 920–926.
19. Ahmadi A, Toroghinejad M R, Najafizadeh A, (2014), 'Evaluation of microstructure and mechanical properties of Al/Al2O3/SiC hybrid composite fabricated by accumulative roll bonding process', *Mater Des*, 53, 13–19.
20. Radhika N, Subramanian R, (2014), 'Effect of ageing time on mechanical properties and tribological behaviour of aluminium hybrid composite', *Int J Mater Res*, 105(9), 875–882.
21. Nataraj M, Balasubramanian K, Palanisamy D, (2018), 'Influence of process parameters on CNC turning of aluminium hybrid metal matrix composites', *Mater Today: Proceed*, 5, 14499–14506.
22. Suresh P, Marimuthu K, Ranganathan S, Rajmohan T, (2014), 'Optimization of machining parameters in turning of Al–SiC–Gr hybrid metal matrix composites using grey-fuzzy algorithm', *Trans Nonferrous Met Soc China*, 24, 2805–2814.
23. Selvakumar V, Muruganandam S, Tamizharasan T, Senthilkumar N, (2016), 'Machinability evaluation of Al–4%Cu–7.5%SiC metal matrix composite by Taguchi–Grey relational analysis and NSGA-II', *Sadhana*, 41, 1219–1234.
24. Radhika N, Subramanian R, (2013), 'Effect of reinforcement on wear behaviour of aluminium hybrid composites', *Tribology*, 7(1), 36–41.
25. Vignesh M, Venkatesan K, Ramanujam R, Vijayan S, (2015), 'Tool wear analysis of Al 6061 reinforced with 10wt% Al2O3 using high hardened inserts', *Appl Mech Mater*, 787, 643–647.
26. Vignesh M, Ramanujam R, Rajyalakshmi G, Bhattacharya S, (2021) Application of grey theory and fuzzy logic to optimize machining parameters of zircon sand reinforced aluminum composites, in *Advanced Fluid* Dynamics, pp. 653–662. Springer, Singapore.
27. Veeresh Kumar G B, Rao C S P, Selvaraj N, (2012) 'Studies on mechanical and dry sliding wear of Al6061–SiC composites', *Compos B Eng*, 43(3), 1185–1191.

7 Erosion Studies on Al/TiC/RHA Reinforced Hybrid Composites through Response Surface Method

K. Balamurugan, Y. Jyothi, and Chinnamahammad Bhasha
VFSTR (Deemed to be University)

S. Vigneshwaran
Kalasalingam Academy of Research and Education

CONTENTS

7.1 INTRODUCTION

The low density, high specific strength materials are required for potential applications due to the composite concept to come into action. Key classifications of composites include metal matrix composites (MMCs), ceramic matrix composites

DOI: 10.1201/9781003345466-7

(CMCs), and polymer-metal matrix composites (PMCs). The metallic properties of traditional alloys are strengthened by the addition of ceramic particles. Lightweight metals have good tribomechanical properties at ambient temperatures, but they diminish at relatively high temperatures. To enhance these properties of metals at ambient and high temperatures, one of the best promising way is the addition of ceramic/agro-industrial waste particles via liquid metallurgy (or) solid route well known as MMCs [1]. The performance of MMCs depends on the selection of matrix, reinforcements, and processing parameters. During the mid-twentieth century, the application of MMCs was increased in various engineering and structural applications. These materials exhibited superior performance than the unreinforced metals. MMCs have enhanced strength, hardness, high strength to weight ratio, superior thermal behavior, and wear resistance characteristics [2]. In MMCs, aluminum-based composites are the competent materials among the present engineering material [3,4]. In recent years, Al-MMCs have been successfully implemented in several engineering disciplines due to lightweight, wear-resistant, and superior mechanical properties i.e., 0.2% proof yield strength, ultimate tensile strength, and hardness[5]. Al is one with the non-ferrous alloys with high resistance due to low weight, superior stiffness, strength, corrosion resistance, machinability, weldability, and high thermal conductivity. Al atoms are arranged in a face-centered-cubic structure with a melting point of 660°C. Al reacts with different elements that developed a series of aluminum alloys (AA). There have been nine series of AA available, four of which are heat-treated AA. This AA has enriched mechanical properties due to precipitation strengthening mechanisms. Series of AA are 1xxx, 2xxx, 3xxx, 4xxx, 5xxx, 6xxx, 7xxx, 8xxx, and 9xxx; among them Al6061 is best suitable for structural applications. Al6061 contains magnesium (1.1%) and silicon (0.4%) as major alloying elements [6]. Al6061 alloy contains moderate strength and poor wear resistance, better weld and wettability, good fatigue strength, and corrosion resistance [7]. Further physical and tribomechanical properties of AA are enhanced in several ways. Al composites were fabricated with different reinforcements (Rf) like SiC [8,9], B_4C [10,11], Al_2O_3 [12,13], TiC [14,15], graphene [16,17], industrial wastes like fly ash, RHA [18,19], and also various nano-based powders [20,21].

The variation of TiC wt. percent i.e., 5%, 10%, and 15% is used in the Al6063/TiC composite manufacture by stir casting. Wear experimentation is carried out. The performance factors were Rf, load, and sliding distance (Sd). The output response was a specific wear rate. Box-Behnken design is implemented to design the Taguchi L15 orthogonal array. The interaction between Rf and Sd significantly affects the specific wear rate. The applied load showed a less significant effect on output responses. The added Rf particles showed excellent wear-resistant than Al6063 [22]. Semegn et al. [23] prepared stir rheocast Al composites with 2%, 4%, and 6% of TiB2. The physical and mechanical properties accelerated up to 4% beyond that, i.e., 6% slow down due to the formation of voids and a bunch of particles. The Rf particles act as nucleation sites and grain refining agents. Bhagya et al. [24] manufacture Al/graphene oxide (GO) composite employing a solid-state process. An increase in wt% of GO enhances the strength and microhardness of Al-MMCs. SEM micrographs display an increase in the Rf content of the formation of minute pores and increased clusters of particles. XRD patterns correspond to the existence of the essential elements. The Rf particles randomly dispersed, and the proper mixing of Rf contributes to the improvement of metallic properties. TiC particle reinforced with Al6061 by stir casting technique to prepare Al-MMCs. The dry wear behavior of Al-MMC worn surface revealed different wear mechanisms. Metallography studies showed that there is a good dispersion of Rf particles, a good bonding nature, and no intermetallic elements [25]. Tribomechanical properties of Al-MMCs depend on reinforcing particle type, shape, size, and weight and volume percent. To strengthen Al synthetic ceramic or agro-industrial waste particulates known as Al-MMCs are added. Various methods are available to develop Al-MMCs among popular, economical, easily controlled casting variables and widely utilized routes i.e., liquid metallurgy. Al hybrid composites (AHCs) showed better physical and tribomechanical properties of single reinforced Al-MMCs [26]. Characterization of the developed AHCs is essential if only Al-MMCs can be replaced.

The mixture of two or more materials forms a novel composite known as AHCs. Al, Mg, Ti, Zn, and Cu as the matrix and synthetic ceramic and agro-industrial particles as reinforcements are used to develop hybrid composites. Among them, AA is widely utilized in various fields of engineering, especially to manufacture several parts in aviation and automobile due to lightweight and high mechanical properties [27]. Al/Al2O3/TiC hybrid composites are developed through the route of the melt. The wt% of Rf was added equally i.e., 5%, 10%, 15%, and 20% respectively. The Rf is preheated to 350°C to remove moisture and improve wetting nature between Al and Rf. The addition of Rf is done in two stages. In the first stage, the preheated Al_2O_3 Rf particles are added to the molten metal, and the stirring operation was carried out with optimum rpm. In the second level, preheated TiC particles are added to the composite slurry as a stirring operation. Characterization data enclosed accelerated hardness, tensile strength, and wear resistance compared to Al1100 alloy [28]. Al6063 reinforced with different wt% of waste particles such as ash of meta cake, straw, and mortar by two-step stir casting is used to develop Al-MMCs and AHCs. The addition of ash leads to decreased density of the composites without compromising specific strength, hardness, and toughness. AHCs showed better mechanical properties than Al-MMCs [29]. B4C and BN as Rf materials developed AHCs using ultrasonic stir casting. The Rf preheated to 450°C leads to enhance wettability while stirring action. Microhardness, the strength of tensile and impact, % of elongation, and metallography are studied. From the observation data, increases in wt% of Rf decrease % of elongation. Mechanical properties diminish after 2 wt% of B4C and BN due to the agglomeration of particles confirmed through micrographs. XRD pattern confirms the presence of principal elements such as Al, B4C, BN, and no intermetallic elements. The addition of particles reduces the size of grains and their boundaries—the enhanced mechanical properties due to the hall-patch strengthening mechanism. The increase in wt% of Rf causes ductile to brittle mode failure during impact load [30]. Al6061 reinforced with TiC and graphite particles via stir casting technique prepared AHCs. TiC, as primary Rf, contributes to enhancing tribomechanical properties. Graphite as secondary Rf is a solid lubricant that increases machinability and sound surface roughness. The graphite wt% increases the tribomechanical properties decelerated. The SEM photographs show the worn surface of

AHCs and enclose several wear mechanisms such as plowing grooves, wear track, delamination, pulled out Rf particles, adhesive, and abrasive wear [31].

Red mud (RM) is one of the world's largest industrial wastage; every year, 150 million RM is produced. RM creates environmental and soil problems. RM and TiC as Rf particles fabricated Al-MMCs using liquid metallurgy and their wear properties are compared. The addition of Rf leads to the formation of a group of fine particles visible from SEM micrographs. XRD analysis confirms the presence of Rf, Al, Fe_2O_3, and no intermetallic element. Wear test is carried out by using a pin on disc setup. TiC is a well-known wear-resistant Rf. The wear rate and friction coefficient of Al/TiC are 19% and 8% lower than Al/RM composites. Al/RM showed better wear properties than Al. The industrial waste showed significant enhancement of wear properties of Al-MMCs [32].

Material researchers have shown interest in preparing composites for the past three decades to meet global requirements due to their high resistance to wear and corrosion eco-friendly, lightweight, and high performance. Hybrid composites are characterized as having a base metal with different types of ceramic reinforcement combined to form a highly promising material with excellent mechanical and tribological properties. The hybrid composite was created by combining the SiC micron and the TiC particles, respectively, by the use of the stir casting. The results show that the mechanical properties decreased due to the particles' agglomeration at higher percentages of weight [33]. The mechanical, physical, electrical, and thermal attributes quickened by adding SiC into the Al metal matrix developed mono and green composites with variable percentage weights of RHA. Mono and bilayer composites are studied by the design of experiments and analysis of variance. Optimal processing conditions are revealed for maximizing bending strength and electrical conductivity. Minimizing the thermal expansion coefficient will lead to have optimal conditions, as researchers had developed Al composite using Agro-waste particle, i.e., RHA [34]. Al composites are developed with waste particles, i.e., RHA. The addition of 18% of RHA particles improved mechanical properties.

The waste particles act as heterogeneous nucleation sites, and grain refiners contribute to enhance metallic properties. Optical studies reveal the homogeneous distribution of RHA particles and excellent bonding with Al matrix [35]. Al composites are fabricated and drilling studies are performed with Taguchi L27 orthogonal array. The influencing parameter was drill bit angle, cutting speed, wt% of reinforcements, and feed rate. Formation of burr on the inner surface of the machined surface have increase the Ra response. The quadratic model is developed by response surface methodology (RSM). Drill bit angle, cutting speed, and wt% of SiC showed a significant effect on responses. Ra decreased, while the force of thrust increased; this leads to an increase in the temperature and the feed rate [36].

Al matrix mixed with ZrO_2 and coconut shell ash (CSA) developed AHCs by the stir casting process. The density, hardness, and bending strength of the AHCs increase by 11.1%, 31.5%, and 9.52% than Al6082. The impact and % of elongation reduce with increases in CSA particles. Increasing wt percent of Rf particles shown marginal improvement of properties occurred due to improper distribution of particles and porosity [37]. Al reinforced with B4C and cow dung ash fabricated AHCs via the stir casting process. The tensile fracture failure patterns were cleavage facets, particle agglomeration, impoverished bonding region, and microvoids and cracks observed via SEM examination. The increase in reinforcement's % leads to form cleavage facets due to a significant reduction in plastic deformation. Cleavage facets are shaped in brittle fracture and tend to initiate microcrack via grains, the region of weak bonding, and voids. The applied load more significant than the specific tensile strength of the casted Al composites leads to initiating microcrack and its propagation [38]. Fabrication of Al composites follows different methods like stir casting, ultrasound stir casting, liquid infiltration, spray deposition, squeeze casting, and powder metallurgy. Of all these techniques casting is the most preferred method because of its efficiency and economics [39]. Manufacturing of Al composites in a large scale has significant problems such as wettability, clustering of particles, level of porosity, and chemical reactivity [40]. Homogeneous distribution of particles, improved wettability, and minimization of porosity levels are achieved by ultrasonic stir casting [41].

The present generation needed low-cost, lightweight materials with higher strength, stiffness, resistance to wear, and corrosion for the budding applications. Al-MMCs had a lower density, high wear resistance, and temperature resistance because Al-MMCs in various engineering fields almost replaced existing alloys. In the third generation, most of the conventional ferrous alloys are replaced with Al-MMCs in potential uses. Due to the unique properties, Al-MMCs are widely used in aerospace, defense, and automobile. Duralcan and Nissan have successfully produced and introduced pistons, brake pads, liners, connecting rods, and propeller shafts [42]. Toyota, Honda, and Martin Marietta successfully manufacture and implement pistons, connecting rods, engine blocks, and piston rings by Al/Al_2O_3 and Al/TiC composites [42]. Using Al-MMCs is successfully developed and implemented in the automobile sector. Al composite's unique and specific properties have the potential for being used in various applications among transport of rail, maritime, aerospace, automobile, and their structural and non-structural parts.

Monolithic and MMCs perform limited applications because of moderate metallic properties. Further, increase in metallic properties is by the addition of nanoparticles for developing nano-AHCs. To develop nano-AHCs various reinforcements are needed like ceramics, industrial/agro waste, CNT, MWCNT, graphine, etc. [43]. Most of the researchers concern various AA series among precipitation strengthened alloys commercially available and a wide variety of applications. Al-based matrix reinforced with rice husk ash and SiC particles is prepared by the solid-state process. 10% of SiC is fixed and RHA wt% varied i.e., 5%, 10%, and 15%. Characterization results reported that the addition of 10% of RHA leads to density of the hybrid composite, yield, tensile properties, and increased fracture toughness.

Limited wt% of RHA reveals better tribomechanical properties than after agglomeration of particles observed. The cost of the green composite reduces due to economic Rf [44]. Al-Mg-Si alloy/rice husk ash/Al_2O_3 is fabricated by double stir casting process; adding rice husk ash observed increases in corrosion rates [45]. A356.2/rice husk ash/SiC is built by the liquid metallurgical route and hardness, yield, and tensile properties increased with increase in reinforcement weight fractions, but % of elongation decreases [46]. Fabricated Al_2O_3 and rice husk ash reinforced with Al6068 matrix by stir casting process confirms that increase in rice husk ash composition may slightly decrease the mechanical properties. However, 2% weight fractions of RHA improve the fracture toughness, specific strength and % elongation than pure alumina. Increasing the weight percent of reinforcement's marginal improvement of properties occurred due to improper distribution of particles and porosity [47]. Hybrid composite is fabricated by stir casting with micron SiC and TiC particles, respectively. Results reveal that mechanical properties reduced because of agglomeration of particles at higher weight percentages. Al nanocomposites are prepared through the combined vertex, stirrer, and ultrasonic cavitation techniques. 1% of SiC is Rf. Due to the presence of a combined effect, nanoparticles have been continuously distributed in the matrix. Micrographs reveal silent features such as new grains formations, clear interface, particle size and distribution, fracture mechanisms, etc. 79% porosity levels decreased than unreinforced Al. Bunch and particle-set splits and uniformly distributed particles confirmed by micrographs lead to accelerated mechanical properties [48]. Vinod et al. developed hybrid composites with fly ash and RHA by using stir casting. The addition of mechanical reinforcement properties improved up to 10% because waste particles act as grain refiners, and proper distribution achieved beyond those mechanical properties decelerated due to a set of particles [49]. Al composite manufacture by stir casting and machining is carried out on it.

Taguchi orthogonal array presented with governing parameters and responses. It is revealed that chip formation continues at higher spindle speed with lower Ra and lower spindle speed with higher Ra. Ra decreased, while the force of thrust increased; this led to an increase in the temperature and the feed rate at the same time. Agro waste particles play an important role in machining. Agro Rf acts as a solid lubricant during machining. The addition of waste Rf particles improved the chip removal rate and reduced Ra [50]. Over the past two decades, the use of AA with different materials as Al-MMCs has tremendous outcomes with the required properties. Employing a sonication method, Al-MMCs have prepared and revealed that ultrasound breaks clustered nanoparticles. Uniform distribution of the particles in the matrix is confirmed by micrographs. Wettability enhances during the stirring process. Porosity levels reduce compared to the base alloy.

Machining the Al-MMCs is fascinating and exciting. Machining of Al composites and AHCs was difficult to achieve closer tolerances due to the presence of hard ceramics. There are traditional and non-conventional methods available. Drawbacks of the conventional process were higher wear of the flank, poor machinability, and unprofitable process. Among milling is a common and essential process of machining to attain closer tolerances [51]. Al matrix reinforced with TiB2 particles has developed Al-MMCs. Composite Al-TiB2 is suitable for aero-engine blades. The presence of hard ceramics restricts budding applications. The machining behavior of Al-TiB2 has been experimentally studied, and the optimal process parameters showed to maximize material removal and minimize Ra. The cutting force is higher than Al due to the presence of TiB2 particles. Developed residual forces are negligible for Al and Al/TiB2 composites [52].

In the same way, composite Al-TiB2 is developed by stir casting technique and mailing carried out experimentally. Cutting zone temperature model is analytically developed and experimental tests confirmed. The established model showed good agreement with the experimental results. The temperature produced in the cutting zone affected Ra, residual stress, and machining defects [53]. Al/TiB2/TiC in-situ composite is prepared through stir casting and experimentally machining behavior investigated. Developed composites are characterized by SEM, EDS, XRD, and microhardness. Rf particles are homogeneously dispersed in the Al matrix. Various levels of parameters such as cutting speed, cutting depth, and feed rate effect Ra, and tool wear examined. The cutting temperature increases due to the presence of Rf, high cutting speed, and forces [54]. Al/15% of CSA composites are produced with stir casting and experimentally machining behavior examined. They were designed by the way of RSM experiments. The spindle speed, feed rate, and cutting depth were process parameters, and response is Ra investigated. The CSA is a soft ceramic that act as a solid lubricant that reduces Ra and tool wear. Feed rate and cutting depth are the most dominant factors on Ra while machining the composite [55]. The addition of ceramics deficit is to cut Al-based composites. The addition of reinforcements generates high temperatures during cutting, which causes defects during the machining process.

The surface integrity of composites is affected by imperfections such as Rf particles peel off, interface debonding, Rf particles pulled out from ductile phase, microcrack, residual stress, thermal fields, and Rf particles compressed. Al-based composites are characterized by SEM with EDS, OM, and XRD patterns [56]. Different wt% of fly ash reinforced with Al matrix prepared A-MMCs through liquid metallurgy. Erosion studies are performed using air-jet erosion tester with varying angles of penetration. SiO_2 was an erosion particle with an average size of 250–500 microns. The wear rate increases due to the rise of wt% of Rf, angle of impact, and velocity. At lower velocity and impact angle erosion mechanisms were microplowing and cutting. At higher velocity and impact angle erosion is due to fracture. SEM micrographs enclosed wear mechanisms [57].

The Al alloy Al6061 used in the present work has high strength at normal environmental conditions, but at high-temperature applications, this alloy fails because of the weak thermal property. This can be solved by the reinforcement of TiC particles [58]. Al6061 is used for automotive to aerospace applications. TiC particles are thermodynamically stable at higher temperatures, have unique properties, and improve wettability. TiC is the attractive reinforcement in the MMCs, which offers increased modulus, strength, high hardness as well excellent wear resistance to the matrix [59–62]. Yet, Al6061/TiC composites have some drawbacks like low toughness and ductility. The reinforcement of secondary reinforcement eliminated these drawbacks by various researchers [63–66].

Authors in their previous work studied the effect of secondary reinforcement in Al6061/TiC composite. They used rice husk char as the secondary reinforcement and witnessed considerable increment in the toughness and ductility of the Al6061/TiC composite. RHC particles are freely available, have better dispersion, good strength, hardness and resistance to wear, superb machinability, and low frictional coefficient. The reinforcements were nanoparticles of TiC, and RHC serves as heterogeneous nucleation sites and grain refiners. The addition of reinforcements enhances the mechanical properties of Al composites due to the dispersion strengthening mechanism. The increased toughness of these hybrid composites could increase the erosion resistance of the composite. These composites experience erosion wear since they were used in applications like turbine materials and dusty industrial applications. So it became essential to study the hybrid composite erosion property, and hence the present work is designed. To the best of the author knowledge, no work has been reported in the solid particle erosion wear behavior of Al6061 based composites.

In the present investigation, the Al6061/TiC/RHC hybrid composites were fabricated through the stir casting process, and solid particle erosion behavior was investigated, and also erosion results were compared with the Al6061 alloy. The erosion experiments were designed and analyzed using RSM. The optimum level of erosion variables was found by using desirability analysis. Finally, the hybrid composites erosion mechanism was studied using scanning electron microscopy images.

7.2 MATERIAL AND METHODS

7.2.1 RAW MATERIALS

The matrix Al6061 aluminum alloy was bought from Coimbatore Metal Mart, India, and its material composition is shown in Table 7.1. The reinforcement titanium carbide (TiC) was obtained from Nano Wings Pvt. Ltd, Hyderabad. The rice husk was collected from the local rice mills in Guntur, AP, India. The rice husk ash was prepared by heating the rice husk at 300°C inside a closed container for 2 hours. After the completion of the heating process, the rice husks were converted into dark black colored char, and these chars were again heated at 610°C for 24 hours for reducing the carbon content in their composition. After 24 hours, the char particles were converted into grayish white ash powder. The obtained ash powder was ball milled using TENCAN to attain uniform size. The photocopy of conversion of rice husk to ash is shown in Figure 7.1 (a–d). The SEM and EDAX analysis results of the particle reinforcement are shown in Figure 7.2.

The two-step ultrasonic-assisted stir casting process was used for the fabrication of the hybrid Al6061 composite. The stir casting set up used for fabrication is shown in Figure 7.3. The ultrasonic transducer working at the power of 2KW produces ultrasonic waves at a maximum of 20 kHz. Initially, the matrix is melted and converted into molten metal at 800°C. The melt starts to cool and changed to semi-solid form.

TABLE 7.1
Elemental Composition of AA6061 Alloy

Elements	Weight % in AA6061
Silicon (Si)	0.4
Magnesium (Mg)	1.1
Copper (Cu)	0.02
Iron (Fe)	0.7
Chromium (Ch)	0.3
Manganese (Mn)	0.07
Zinc (Zn)	0.15
Aluminum (Al)	Remaining

FIGURE 7.1   Various stages of conversion from rich husk to rice husk ash. (a) Raw riice husk, (b) Rice husk char, (c) Sieved RHC, (d) Heat treated RHC.

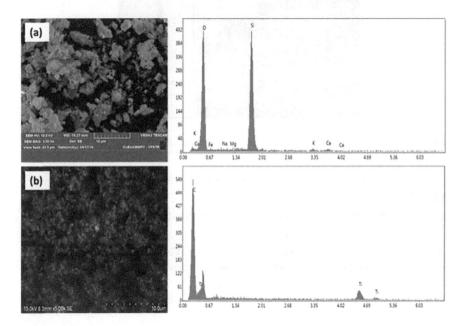

FIGURE 7.2   SEM and EDAX analysis of particle reinforced: (a) RHA and (b) TiC.

The semi-solid melt was then stirred. To increase the matrix wettability, 1% of magnesium was added, and again the preheated reinforcements were added to the melt. To avoid the oxidation of materials, the melting and mixing process was carried out in the inert environment. At this stage, the ultrasonic probe is introduced to the melt mixture. The ultrasonic treatment ensures the proper dispersion of the reinforcing particles without any vortex formation. Then the melt was poured into the preheated mold (450°C). The same process was repeated for fabricating hybrid Al6061 composite at varying reinforcement weight percentage. Eventually, hybrid composite test ingots were developed in the sizes of $300 \times 120 \times 15\,mm^3$. Samples are machined using abrasive water jet machining (AWJM) to the dimensions of $25 \times 25 \times 10\,mm^3$. The properties and details of the fabricated Al6061 hybrid composites are shown in Table 7.2.

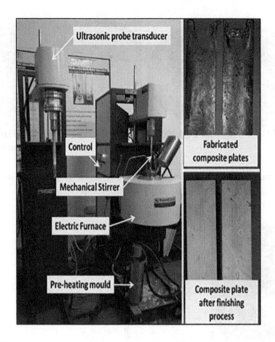

FIGURE 7.3   Fabrication setup and composite fabricated.

TABLE 7.2
Mechanical Properties of Al6061 Hybrid Composite

Material Combination	Specification	Tensile Strength (MPa)	Flexural Strength (MPa)	Vickers Hardness	Density (g/cm³)	Void Percentage (%)
AA6061 pure	Sample 1	119	112	66	2.68	0.38
Hybrid AA6061 Composite (3% RHA and 3% TiC)	Sample 2	134	115	78.28	2.72	1.26
Hybrid AA6061 Composite (3% RHA and 6% TiC)	Sample 3	202	219	89.13	2.8	0.67
Hybrid AA6061 Composite (3% RHA and 6% TiC)	Sample 4	183	176	83.47	2.83	1.82
Hybrid AA6061 Composite (3% RHA and 6% TiC)	Sample 5	156	170	81.65	2.84	3.78

7.2.2   Solid Particle Erosion Test

In the solid particle erosion test, the hard abrasive particles are targeted on the sample material at high velocity. The repeated impact of the abrasive particles induces material loss because of erosion wear. The erosion wear in the material was calculated in terms of erosion rate using the expression 1 [67]. The expression uses the weight of the sample before and after the erosion test, erodent flow rate, and the time of sample exposure to erosion.

$$\text{Erosion rate} = \frac{\text{Weightloss}}{\text{discharge} \times \text{time}}\ (g/g) \tag{7.1}$$

The erosion test rig supplied by the DUCOM, India, was used for conducting the erosion experiment. The test rig consists of a compressor with supply dry compressed air. The abrasive particles get mixed with the dry compressed air and passed through the nozzle at high velocity and made to impact on the sample placed at the holder. The angle of abrasive was varied by changing the position of the sample using a different specimen holder. The abrasive particle velocity was varied by controlling the air pressure. Before and after the erosion test, samples were well cleaned by acetone and dry compressed air to avoid the presence of dust and abrasives. Each sample was tested thrice, and the average value of weight loss has been used to find the erosion rate.

7.2.3   Design of Experiment Based on RSM

The experiment table for investigating the erosion rate of the hybrid Al6061 composites was designed through RSM-central composite design (CCD). RSM is the statistical tool; it consists of mathematical calculations for analyzing and optimizing the input and output variables. RSM involves the development of a functional relationship between the input response and the output response. The main reason for using the RSM method is that the relationship between the input and output variables can be found. The effect of variation of each parameter on the output variable can be clearly understood. It is also used to find the combination of input variables in producing the optimum output value. In the present work, using RSM a mathematical model was developed to investigate the characteristics of the erosion rate of hybrid Al6061 composite. The experimental erosion variables were coded with the maximum, intermediate, and minimum values 1, 0, and −1. The erosion variables and their level are shown in Table 7.3. The experiment matrix developed with the

TABLE 7.3
Experimental Parameters and Their Coding

Coded Level	TiC Reinforcement (%)	Impact Angle (°)	Erodent Velocity (m/s)
−1	0	30	100
0	3	60	125
+1	6	90	150

TABLE 7.4
Experiment Design and Results

S. No.	Std	Erosion Variables			Erosion Rate $\times 10^{-6}$ (g/g)
		TiC Reinforcement (%)	Impact Angle (°)	Erodent Velocity (m/s)	
1	15	0	0	0	12.36
2	14	0	0	1	13.16
3	2	1	−1	−1	11.12
4	1	−1	−1	−1	16.43
5	12	0	1	0	10.49
6	5	−1	−1	1	19.41
7	11	0	−1	0	13.99
8	13	0	0	−1	10.92
9	17	0	0	0	11.94
10	9	−1	0	0	14.21
11	3	−1	1	−1	10.21
12	4	1	1	−1	7.92
13	8	1	1	1	10.93
14	18	0	0	0	12.92
15	16	0	0	0	13.02
16	10	1	0	0	11.22
17	7	−1	1	1	12.49
18	6	1	−1	1	13.28

coded and erosion rate of the corresponding experiment is shown in Table 7.4. The experimental matrix consists of 18 sets of coded conditions designed with the α value 1 and center point 4.

7.3 RESULTS AND DISCUSSION

7.3.1 REGRESSION MODELING

To study the effect of erosion variables, TiC reinforcement % (A), impact angle (B), and erodent velocity (C) on the erosion rate (Y), a second-order polynomial regression model was formulated using the expression,

$$y = \beta_0 + \sum_{i=1}^{k} \beta_i X_i + \sum_{i,\,j=1}^{k.} \beta_{ij} X_i X_j \, [[\text{Tab}]][[\text{Tab}]] \tag{7.2}$$

where β is the regression coefficients, y is the output variable, x is the input variable, and k is the total number of input variables.

The second-order response equation for the erosion rate (Y) of Al 6061 hybrid composite can be expressed as the function of A, B, and C as follows:

$$Y = \beta_0 + \beta_1 A + \beta_2 B + \beta_3 C + \beta_{12} AB + \beta_{13} AC + \beta_{23} BC \left[[\text{Tab}]\right] \qquad (7.3)$$

Using the coded variables and actual variables, the final mathematical equation was found using the design expert software package. This final expression developed to predict the erosion rate of the developed Al6061 hybrid composite is,

$$ER = 0.125567 - 0.018280 * A - 0.022190 * B + 0.012670 * C + 0.009488 * AB$$

$$-0.000112 * AC + 0.000187 * BC \qquad (7.4)$$

The ANOVA analysis was performed to evaluate the developed model adequacy. The ANOVA results of the developed linear model are shown in Table 7.5. The model adequacy is the measure of R^2, adjusted R^2, and predicted R^2 values. The R^2 value defines the model goodness, which explains the closeness of the data with the developed regression model. Adjusted R^2 is the variation in the R^2 that adjusts the variables in the model. In the present model, the R^2 and adjusted R^2 values are above 95%, which means they are close to 1. This indicates the agreement of the developed model. Besides, the predicted R^2 value (88.43%) describes the fitness of the fitted model. This indicates that the response variable can be predicted with the desired experimental variable range through the regression model.

The significance of the model and erosion variables can be understood with the P-value and F-value. F-value signifies the influencing level of the corresponding factor on response. Higher F-value signifies higher influence. ANOVA analysis was done at a 95% confidence level (P-value 0.05). P-value < 0.05 explains the significance of the factor on response. The model F-value is 65.71, and P-value is 0.0001, and this indicates the developed model is significant. All three variables, TiC reinforcement,

TABLE 7.5
ANOVA for Erosion Rate

Source	Sum of Squares	df	Mean Square	F-Value	P-Value	Contribution %
Model	105.91	6	17.65	65.71	< 0.0001	
A-TiC Reinforcement	33.42	1	33.42	124.39	< 0.0001	29.88
B-Impact angle	49.24	1	49.24	183.29	< 0.0001	44.03
C-Erodent velocity	16.05	1	16.05	59.76	< 0.0001	14.35
AB	7.20	1	7.20	26.81	0.0003	6.44
AC	0.0010	1	0.0010	0.0038	0.9521	0.00
BC	0.0028	1	0.0028	0.0105	0.9203	0.00
Residual	2.96	11	0.2686			2.65
Lack of Fit	2.19	8	0.2737	1.07	0.5315	1.96
Pure Error	0.7656	3	0.2552			0.68
Cor Total	108.87	17				

R^2 97.29%; Adjusted R^2 95.80%; Predicted R^2 88.43%.

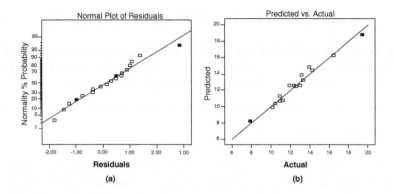

FIGURE 7.4 (a) Probability plots. (b) Comparison between experimental results and predicted values of erosion rate of composites.

impact angle, and erodent variable, are found to be significant. The impact angle has the F-value of 183.29 and contribution percentage 44.03%, which implies the higher significance of impact angle in affecting the erosion rate of the Al6061 hybrid composite. Next to impact angle, TiC reinforcement shows 29.88% of contribution in affecting the composite erosion rate followed by erodent velocity with 14.35% contribution. In interaction, TiC reinforcement and impact angle significantly affected the erosion rate; this is evident in the interrelation between these two parameters.

Figure 7.4a shows the normal probability plot, and Figure 7.4b shows the variations in the predicted data. The normal distribution of the predicted and experimental values can be understood with the normal probability plot, which directs the formulated model's significance. In the probability plot, the data points lie close to the straight line indicating the normal distribution of error value and acceptance of the fitted model. In Figure 7.4b, the correlation of the predicted values with the experimental data values can be understood. There is no deviation between the predicted and experimental data; predicted data fit well with the experimental data.

7.3.2 Erosion Wear Mechanisms

Figure 7.5 shows the effect of erosion variables and TiC influence against the erosion rate of the Al6061 hybrid composite. The erosion rate is maximum at the pure Al6061 alloy. The increased ductility of the alloy material is the main reason for the high erosion rate. The addition of TiC increased the Al6061 matrix erosion resistance property. The erosion rate is lower on 6% TiC-reinforced composites. The addition of TiC reduced the composite ductile nature and increased surface hardness.

This offered greater resistance against the erodent particle impact. All the composites showed a maximum erosion rate at 30° impact angle. At 30° the erodent particle easily skids over the metal surface, repeated skidding of erodent particle plow the surface, and removes the material. The micrograph (Figure 7.6) images of the eroded specimen at 30° and 60° show the presence of plowed surface with microcuts and lip formation, which confirm the ductile erosion. Unlike at 30° and 60°, at 90° there is no

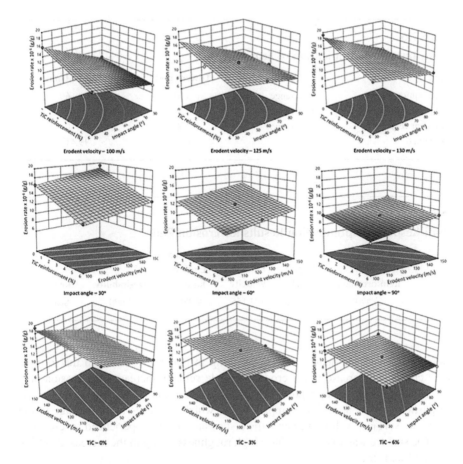

FIGURE 7.5 Effect of parameters on erosion rate.

(a) Impact angle (30°)   (b) Impact angle (60°)

FIGURE 7.6 Erosion test sample with 6%, TiC-reinforced composite.

chance for the erodent particle to slide on the composite surface. Instead, the erodent impacts on the surface and produce strain hardening due to repeated attack. Under strain hardening, the material undergoes plastic deformation. But on the continuous impact of erodent particle, the surface develops microcracks. The cracks branch and

TABLE 7.6

Constraints for Optimization of Erosion Rate

Name	Goal	Lower Limit	Upper Limit
A: TiC reinforcement	is in range	0	6
B: Impact angle	is in range	30	90
C: Erodent velocity	is in range	100	150
Erosion rate	none	7.92	19.41

TABLE 7.7

Multi-Response Optimization Results for Erosion Rate

S. No.	TiC Reinforcement	Impact Angle	Erodent Velocity	Predicted Erosion Rate of	Desirability	Remarks
1	3.470	85.658	103.155	9.380	1.000	Selected

interconnect with the nearby cracks, resulting in material removal in the form of small chips. An increase in the erodent particle velocity increased the erosion rate. This trend was similar, as reported by previous literature. At high erodent velocity, the particle strikes at a higher impact force, which produces more damage to the composite surface. At high velocity, the composite surface experienced severe damage. The surface was found to behave high roughness owing to the increased cutting tears and wear traces.

7.3.3 OPTIMIZATION OF EROSION RATE USING DESIRABILITY ANALYSIS

The optimum range of erosion variables in producing minimum erosion rate on Al6061 hybrid composites is analyzed with multi-response desirability analysis. A desirability function is developed in desirability analysis using the input and output response. Table 7.6 shows the condition used for the desirability analysis of the present work, and Table 7.7 shows the results of the analysis. The erosion rate of the Al6061 hybrid composites can be minimized at the condition of 3.470 wt% reinforcement of TiC, impact angle of 85.658°, and erodent velocity of 103.155 m/s. The desirability value 1 indicates that all the input responses are within the range.

7.4 CONCLUSIONS

In the present investigation, applications of RSM and CCD for modeling the erosion rate of TiC/RHC reinforced Al6061 hybrid composite are presented. Through regression modeling the second-order polynomial mathematical model was developed as per CCD. The model was developed by considering impact angle, erodent velocity, and wt% of TiC reinforcement. ANOVA analysis on the RSM model evident the model adequacy at a 95% confidence interval. The erosion variables were optimized using desirability analysis. From analysis, the following conclusion can be drawn:

- CCD is found to be an effective method for modeling and predicting erosion wear.
- ANOVA analysis shows that the impact angle and TiC reinforcement percentage are two dominant factors that contributed to affecting the erosion variable.
- The optimum set of selected erosion variables are impact angle of 85.658°, TiC reinforcement of 3.470 wt%, and erodent velocity of 103.155 m/s, which results in a minimum erosion rate of 9.380 g/g.
- The eroded surface showed ductile failure at a lower impact angle. At 90° impact angle, the composite surface experiences plastic deformation. Microcutting, cracks, and plowing were found to be the reason for material removal under erosion.

ACKNOWLEDGMENT

The authors are grateful to the Centre of Excellence at VFSTR (Deemed to be University), Guntur, AP, India, for rendering their support and guidance to finish this work.

DECLARATION OF CONFLICTING INTERESTS

The author(s) declared no potential conflicts of interest concerning the research, authorship, and/or publication of this article.

FUNDING DETAILS

This work was supported by VFSTR (Deemed to be University) Guntur-522213, AP, India, under Seed Grant F.No.:VFSTR/Reg/A4/30/2019–20/02 dated 17.07.2019.

REFERENCES

1. Idusuyi N, Olayinka JI. Dry sliding wear characteristics of aluminium metal matrix composites: A brief overview. *Journal of Materials Research and Technology* 8(3) (2019) 3338–3346. https://doi.org/10.1016/j.jmrt.2019.04.017.
2. Clyne TW, Withers PJ. An introduction to metal matrix composites (1993).
3. Mahesh Kumar V. A comprehensive review on material selection, processing, characterization, and applications of aluminium metal matrix composites. *Materials Research Express* 6 (2019) 072001. https://iopscience.iop.org/article/10.1088/2053-1591/ab0ee3/meta (accessed April 23, 2020).
4. Surappa MK. Aluminium matrix composites: Challenges and opportunities. *Sadhana* 28 (2003) 319–334. doi:10.1007/BF02717141.

5. Suthar J, Patel KM. Processing issues, machining, and applications of aluminum metal matrix composites. *Materials and Manufacturing Processes* 33(5) (2017) 499–527. https://doi.org/10.1080/10426914.2017.1401713.

6. Rajeshkumar G, Harikrishna AM, Ajithkumar S. A comprehensive review on manufacturing methods and characterization of Al6061 composites. *Materials Today: Proceedings* 22 (2020) 2597–2605. https://doi.org/10.1016/j.matpr.2020.03.390.

7. Rao G, Vundavilli PR, Meera Saheb K. Microstructural and mechanical behaviour of Al6061/Gr/WC hybrid metal matrix composite. In: Li L, Pratihar D, Chakrabarty S, Mishra P. (eds) *Advances in Materials and Manufacturing Engineering.* Lecture Notes in Mechanical Engineering Springer, Singapore (2020). doi: 10.1007/978-981-15-1307-7_59.

8. Nair SV, Tien JK, Bates RC. Sic-reinforced aluminium metal matrix composites. *International Metals Reviews* 30 (1985) 275–290. doi:10.1179/imtr.1985.30.1.275.

9. Melaibari A, Fathy A, Mansouri M, Eltaher MA. Experimental and numerical investigation on strengthening mechanisms of nanostructured Al-SiC composites. *Journal of Alloys and Compounds* 774 (2018) 1123–1132.

10. Kumar SS, Uthayakumar M, Kumaran ST, Parameswaran P. Electrical discharge machining of Al(6351)-SiC-B4C hybrid composite. *Materials, and Manufacturing Processes* 29 (2014) 1395–1400. doi:10.1080/10426914.2014.952024.

11. Rajkumar SE, Palanikumar K, Kasiviswanathan P. Influence of mica particles as secondary reinforcement on the mechanical and wear properties of Al/B4C/mica composites. *Materials Express* 9(4) (2019) 299–309. doi: 10.1166/mex.2019.1497.

12. Kok M. Production and mechanical properties of Al2O3 particle-reinforced 2024 aluminium alloy composites, *Journal of Materials Processing Technology* 161 (2005) 381–387. doi:10.1016/j.jmatprotec.2004.07.068.

13. Junus S, Zulfia A. Development of seamless pipe based on Al/Al2O3 composite produced by stir casting and centrifugal casting. *Materials Science Forum* 857 (2016) 179–182. doi: 10.4028/www.scientific.net/MSF.857.179.

14. Pandey U, Purohit R, Agarwal P, Dhakad SK, Rana RS. Effect of TiC particles on the mechanical properties of aluminium alloy metal matrix composites (MMCs). *Materials Today: Proceedings, Elsevier Ltd* 2017 5452–5460. doi:10.1016/j.matpr.2017.05.057.

15. D'Brass S, Ravi KR, Nampoothiri J, et al. The effect of melt ultrasound treatment on the microstructure and age hardenability of Al-4 Wt Pct Cu/TiC composite. *Metallurgical and Materials Transactions* B50 (2019) 2557–2565. doi: 10.1007/s11663-019-01683-0.

16. Fadavi Boostani A, Tahamtan S, Jiang ZY, Wei D, Yazdani S, Azari Khosroshahi R, Taherzadeh Mousavian R, Xu J, Zhang X, Gong D. Enhanced tensile properties of aluminium matrix composites reinforced with graphene encapsulated SiC nanoparticles. *Composites Part A: Applied Science and Manufacturing* 68 (2015) 155–163. doi:10.1016/j.compositesa.2014.10.010.

17. Natrayan L, Yogeshwaran S, Yuvaraj L, Kumar MS. Effect of graphene reinforcement on mechanical and microstructure behavior of AA8030/graphene composites fabricated by stir casting technique. *International conference on inventive material science applications : ICIMA* 2019, (2019) https://doi.org/10.1063/1.5131599.

18. Dinaharan I, Nelson R, Vijay SJ, Akinlabi ET. Microstructure and wear characterization of aluminum matrix composites reinforced with industrial waste fly ash particulates synthesized by friction stir processing. *Materials Characterization* 118 (2016) 149–158. doi:10.1016/j.matchar.2016.05.017.

19. Virkunwar AK, Ghosh S, Basak R. Wear characteristics optimization of Al6061-Rice husk ash metal matrix composite using Taguchi method. *Materials Today: Proceedings* (2019) https://doi.org/10.1016/j.matpr.2019.07.731.

20. George R, Kashyap KT, Rahul R, Yamdagni S. Strengthening in carbon nanotube/aluminium (CNT/Al) composites. *Scripta Materialia* 53 (2005) 1159–1163. doi:10.1016/j.scriptamat.2005.07.022.

21. Amouri K, Kazemi S, Momeni A, Kazazi M. Microstructure and mechanical properties of Al-nano/micro SiC composites produced by stir casting technique. *Materials Science and Engineering*: A 674 (2016) 569–578. https://doi.org/10.1016/j.msea.2016.08.027.

22. Reddy PV, Prasad PR, Krishnudu DM, et al. An investigation on mechanical and wear characteristics of Al 6063/TiC metal matrix composites using RSM. *Journal of Bio-and Tribo-Corrosion* 5 90 (2019). doi: 10.1007/s40735-019-0282-0.

23. Cheneke S, Karunakar DB. Microstructure characterization and evaluation of mechanical properties of stir rheocast AA2024/TiB2 composite. *Journal of Composite Materials* (2019) 002199831987169. https://doi.org/10.1177%2F0021998319871693.

24. Dasari BL, Morshed M, Nouri JM, Brabazon D, Naher S. Mechanical properties of graphene oxide reinforced aluminium matrix composites. *Composites Part B: Engineering* 145 (2018) 136–144. https://doi.org/10.1016/j.compositesb.2018.03.022.

25. David Raja Selvam J, Dinaharan I, Rai RS, Mashinini PM. Dry sliding wear behaviour of in-situ fabricated TiC particulate reinforced AA6061 aluminium alloy. *Tribology - Materials, Surfaces & Interfaces* (2018) 1–11. https://doi.org/10.1080/17515831.2018.15 50971.

26. Shaikh MBN, Raja S, Ahmed M, Zubair M, Khan A, Ali M. Rice Husk Ash reinforced Aluminium Matrix Composites: Fabrication, characterization, statistical AnalysisAnalysis and artificial neural network modeling. *Materials Research Express* 6(5) (2019) 1–37. doi: 10.1088/2053-1591/aafbe2.

27. Bodunrin MO, Alaneme KK, Chown LH. Aluminium matrix hybrid composites: A review of reinforcement philosophies; mechanical, corrosion and tribological characteristics. *Journal of Materials Research and Technology* 4(4) (2015) 434–445. https://doi.org/10.1016/j.jmrt.2015.05.003.

28. Jamwal UK, Vates PG, Aggarwal A, Sharma BP. Fabrication and characterization of Al2O3-TiC-reinforced aluminum matrix composites. In: Shanker K, et al. (eds), *Advances in Industrial and Production Engineering, Lecture Notes in Mechanical Engineering*. https://doi.org/10.1007/978-981-13-6412-933.

29. Rozhbiany, FAR, Jalal SR. Reinforcement and processing on the machinability and mechanical properties of aluminum matrix composites. *Journal of Materials Research and Technology* 8(5) (2019) 4766–4777. doi:10.1016/j.jmrt.2019.08.023.

30. Dirisenapu G, Pichi Reddy S, Dumpala L. The effect of B4C and BN nanoparticles on the mechanical and microstructural properties of Al7010 hybrid metal matrix nanocomposites. *Materials Research Express* 6(10) (2019) 1–22. https://doi.org/10.1088/2053-1591/ab3d6d.

31. Jeyaprakasam S, Venkatachalam R, Velmurugan C. Experimental investigations on the influence of tic/graphite reinforcement in wear behavior of Al 6061hybrid composites. *Surface Review and Letters* (2018) 1850173. doi:10.1142/s0218625x18501731.

32. Kar C, Surekha B. Effect of red mud and TiC on friction and wear characteristics of Al 7075 metal matrix composites. *Australian Journal of Mechanical Engineering* (2019) 1–10. https://doi.org/10.1080/14484846.2019.1651138.

33. Sambathkumar M, Navaneethakrishnan P, Ponappa K, Sasikumar KSK. Mechanical and corrosion behavior of Al7075 (hybrid) metal matrix composites by two step stir casting process. *Latin American Journal of Solids and Structures* 14(2) (2017) 243–255.doi:10.1590/1679-78253132.

34. Bahrami A, Soltani N, Soltani S, Pech-Canul MI, Gonzalez LA, Gutierrez CA, Gurlo A. Mechanical, thermal and electrical properties of monolayer and bilayer graded Al/SiC/rice husk ash (RHA) composite. *Journal of Alloys and Compounds* 699 (2017) 308–322. doi:10.1016/j.jallcom.2016.12.339.

35. Dinaharan I, Kalaiselvan K, Murugan N. Influence of rice husk ash particles on microstructure and tensile behavior of AA6061 aluminum matrix composites produced using friction stir processing. *Composites Communications* 3 (2017) 42–46. doi:10.1016/j.coco.2017.02.001.

36. Thakre AA, Soni S. Modeling of burr size in drilling of aluminum silicon carbide composites using response surface methodology. *Engineering Science and Technology, an International Journal* 19(3) (2016) 1199–1205. doi:10.1016/j.jestch.2016.02.007.

37. Ravi Kumar K, Pridhar T, Sree Balaji VS. Mechanical properties and characterization of zirconium oxide (ZrO 2) and coconut shell ash(CSA) reinforced aluminium (Al 6082) matrix hybrid composite. *Journal of Alloys and Compounds* 765 (2018) 171–179. doi:10.1016/j.jallcom.2018.06.177.

38. Manikandan R, Arjunan TV, Nath OP. Studies on micro structural characteristics, mechanical and tribological behaviours of boron carbide and cow dung ash reinforced aluminium (Al 7075) hybrid metal matrix composite. *Composites Part B: Engineering* 183 (2019) 107668. doi:10.1016/j.compositesb.2019.107668.

39. Yashpal S, Jawalkar CS, Verma AS, Suri NM. Fabrication of aluminium metal matrix composites with particulate reinforcement: A review. *Materials Today: Proceedings, Elsevier Ltd* (2017) 2927–2936. doi:10.1016/j.matpr.2017.02.174.

40. Bhoi NK, Singh H, Pratap S. Developments in the aluminum metal matrix composites reinforced by micro/nano particles – A review. *Journal of Composite Materials* (2019) 002199831986530.doi:10.1177/0021998319865307.

41. Wang XJ, Wang NZ, Wang LY, Hu XS, Wu K, Wang YQ, Huang YD. Processing, microstructure and mechanical properties of micro-SiC particles reinforced magnesium matrix composites fabricated by stir casting assisted by ultrasonic treatment processing. *Materials & Design* 57 (2014) 638–645. doi:10.1016/j.matdes.2014.01.022.

42. Shinde DM, Sahoo P, Davim JP. Tribological characterization of particulate-reinforced aluminum metal matrix nanocomposites: A review. *Advanced Composites Letters* 29 (2020) 2633366X2092140.doi:10.1177/2633366x20921403.

43. Casati R, Vedani M. Metal matrix composites reinforced by nano-particles—A review. *Metals* 4(1) (2014) 65–83.https://doi.org/10.3390/met4010065.

44. Shaikh MBN, Arif S, Waseem A. Microstructural, mechanical and tribological behaviour of powder metallurgy processed SiC and RHA reinforced Al-based composites. *Surfaces and Interfaces*15 (2019) 166–179. https://doi.org/10.1016/j.surfin.2019.03.002.

45. Alaneme KK, Apata OP. Corrosion and wear behaviour of rice husk ash—Alumina reinforced Al–Mg–Si alloy matrix hybrid composites. *Journal of Materials Research and Technology* 2(2) (2013) 188–194. doi:10.1016/j.jmrt.2013.02.005.

46. Prasad DS, Shoba C. Experimental evaluation onto the damping behavior of Al/SiC/ RHA hybrid composites. *Journal of Materials Research and Technology* 5(2) (2016) 123–130. doi:10.1016/j.jmrt.2015.08.001.

47. Alaneme KK, Akintunde IB, Olubambi PA, Adewale TM. Fabrication characteristics and mechanical behaviour of rice husk ash – Alumina reinforced Al-Mg-Si alloy matrix hybrid composites. *Journal of Materials Research and Technology* 2(1) (2013) 60–67. doi:10.1016/j.jmrt.2013.03.012.

48. Madhukar P, Selvaraj N, Gujjala R, Rao CSP. Production of high performance AA7150– 1% SiC nanocomposite by novel fabrication process of ultrasonication assisted stir casting. *Ultrasonics Sonochemistry* (2019) 104665. doi:10.1016/j.ultsonch.2019.104665.

49. Vinod B, Ramanathan S, Ananthi V, Selvakumar N. Fabrication and characterization of organic and in-organic reinforced A356 aluminium matrix hybrid composite by improved double-stir casting. *Silicon* (2018) doi:10.1007/s12633-018-9881-5.

50. Madhukar P, Selvaraj N, Rao C. Manufacturing of aluminium nano hybrid composites: A state of review. *IOP Conference Series: Materials Science and Engineering* 149 (2016) 012114. doi:10.1088/1757–899x/149/1/012114.

51. Li J, Laghari RA. A review on machining and optimization of particle-reinforced metal matrix composites. *The International Journal of Advanced Manufacturing Technology* 100 (2019) 2929–2943. https://doi.org/10.1007/s00170-018-2837-5.

52. Jiang R, Chen X, Ge R, Wang W, Song G. Influence of TiB 2 particles on machinability and machining parameter optimization of TiB 2/Al MMCs. *Chinese Journal of Aeronautics* 31(1) (2018) 187–196. doi:10.1016/j.cja.2017.03.012.

53. Xiong Y, Wang W, Jiang R, Lin K. Study on cutting temperature modeling of machined workpiece in end milling in-situ TiB2/7050Al MMCs. *Heat Transfer and Thermal Engineering* 8A (2018) doi:10.1115/imece2018-87319.

54. Kakaravada I, Mahamani A, Pandurangadu V. Investigation on cutting force, flank wear, and surface roughness in machining of the A356-TiB2/TiC in-situ composites. *International Journal of Materials Forming and Machining Processes* 5(2) (2018) 45–77. doi:10.4018/ijmfmp.2018070104.

55. Sivasankara Raju R, Rao CJ, Sreeramulu D, Prasad K. Evaluation of optimization parametric condition during machining for Al-CSA composite using response surface methodology. In: Deepak B, Parhi D, Jena P. (eds) *Innovative Product Design and Intelligent Manufacturing Systems. Lecture Notes in Mechanical Engineering.* Springer, Singapore (2020).

56. Liao Z, Abdelhafeez A, Li H, Yang Y, Diaz OG, Axinte D. State-of-the-art of surface integrity in machining of metal matrix composites. *International Journal of Machine Tools and Manufacture* (2019) doi:10.1016/j.ijmachtools.2019.05.006.

57. Tanusree Bera K, Acharya SK. Solid particle erosion behaviour of cenosphere reinforced LM6 (Al-Si12) matrix alloy composites. *International Journal of Materials Engineering Innovation* 8(3/4) (2017).

58. Kishore DSC, Rao KP, Mahamani A. investigation of cutting force, surface roughness and flank wear in turning of in-situ Al6061-TiC metal matrix composite. *Procedia Materials Science* 6 (2014) 1040–1050. doi:10.1016/j.mspro.2014.07.175.

59. Das K, Bandyopadhyay TK, Das S. A review on the various synthesis routes of TiC reinforced ferrous based composites. *Journal of Materials Science* 37 (2002) 3881–3892. doi:10.1023/A:1019699205003.

60. Parashivamurthy KI, Kumar RK, Seetharamu S, Chandrasekharaiah MN. Review on TiC reinforced steel composites. *Journal of Materials Science* 36 (2001) 4519–4530. doi:10.1023/A:1017947206490.

61. Tyagi R. Synthesis and tribological characterization of in situ cast Al-TiC composites. *Wear, Elsevier* (2005) 569–576. doi:10.1016/j.wear.2005.01.051.

62. Tong XC, Ghosh AK. Fabrication of in situ TiC reinforced aluminum matrix composites. *Journal of Materials Science* 36 (2001) 4059–4069. doi:10.1023/A:1017946927566.

63. Kumar V, Gupta RD, Batra NK, Comparison of mechanical properties and effect of sliding velocity on wear properties of al 6061, Mg 4%, Fly Ash and Al 6061, Mg 4%, Graphite 4%, Fly Ash Hybrid Metal Matrix Composite. *Procedia Materials Science* 6 (2014) 1365–1375. doi:10.1016/j.mspro.2014.07.116.

64. Hima Gireesh C, Durga Prasad K., Ramji K. Experimental investigation on mechanical properties of an Al6061 hybrid metal matrix composite. *Journal of Composites Science* 2 (2018) 49. doi:10.3390/jcs2030049.

65. Umanath K, Palanikumar K, Selvamani ST. Analysis of dry sliding wear behaviour of Al6061/SiC/Al2O 3 hybrid metal matrix composites. *Composites Part B: Engineering* 53 (2013) 159–168. doi:10.1016/j.compositesb.2013.04.051.

66. Chinnamahammad Bhasha A, Balamurugan K. Fabrication and property evaluation of Al 6061 + x% (RHA + TiC) hybrid metal matrix composite. *SN Applied Sciences* 1 (2019) 1–9. doi:10.1007/s42452-019-1016-0.

67. Vigneshwaran S, Uthayakumar M, Arumugaprabu V. Solid particle erosion study on redmud - An industrial waste reinforced sisal/polyester hybrid composite. *Materials Research Express* 6 (2019). doi:10.1088/2053-1591/ab0a44.

8 Machining Studies on AlSi₇+63% SiC Composite Using Machine Learning Technique

K. Balamurugan
VFSTR (Deemed to be University)

T. P. Latchoumi
SRM Institute of Science and Technology

Shivaprasad Satla
Malla Reddy Engineering College (A)

CONTENTS

8.1 INTRODUCTION

The metallic matrix composite (MMC) features a wide range of scales and microstructures. Not unusual to all of them is a continuous metallic matrix. MMCs may be described as the metals which might be strengthened with fibers or particulates that typically are stiffer, stronger, and lightweight. One of the primary functions for enhancing the metals is to improve the stiffness and constituents turned into to increase their stiffness/density and power/density ratios. The normal matrix materials are aluminum, magnesium, titanium, and copper. The main purposes of the infiltration in a matrix are to provide an efficient transfer of load to the reinforcement and to resist cracking in the event of a reinforcement failure; therefore, high strength capacity and excessive strength of the metal matrix should be selected. Reinforcements in MMCs are the second segment that contributes to a matrix alloy, usually leading to a few changes in source properties typically with an increase in hardness, modulus, and strength of the composite materials.

Usually, ceramic particles are used, and the reinforcement particles in MMCs are SiC, Al_2O_3, TiB2, and B4C that are characterized with the aid of their excessive property and stiffness factors both at room and high temperature operating conditions. Sometimes refractory materials inclusive of tungsten and Ti were also used as reinforcements which may find in shape to be for ordinary engineering applications. The main cause of the reinforcement was to reinforce and stiffen the composite by preventing matrix deformation with the aid of mechanical restraint. Normally, reinforcement will increase the toughness, hardness, and the high-temperature functionality of MMCs [1].

Composite substances are flexible in element choice so that the properties of the substances may be tailor-made. The most important drawback of metal matrix composites commonly lies within the production or the manufacturing cost involved and price factor involved in the purchase of the reinforcement substances. Therefore, the cost-effective handling of composite substances is a basic component of the development of their applications. The flexibility of a wide scope of support materials and the advancement of recent preparing methods are important to composites [2]. This is the positive impact of the high-performance aluminum substances, not only because of the characteristics of the composites but also it was an effective approach to the reinforcement of aluminum alloys.

The disintegration of Li into Mg is a minor solution for strengthening purposes and was used to enhance the effect of reinforcement without the formation of any intermetallic phase elements like Mg-Li at the time of losing the heat at the cooling process [3]. For this reason, heat treatment mainly based on phase formation will not show any sort of improvement in the properties of the composite materials. Efforts to improve the strength of this binary system by using LiX (X = Al, Zn, Cd, etc.) were because such precipitates usually appear to overage easily even at room temperature, form precipitates, and have no longer been a success. Alternatively, the incorporation of thermally strong reinforcements into composites makes them satisfactorily acknowledged for excessive-temperature packages. Within the automotive manufacturing sector, the potential of the aluminum matrix composites is their use of disk rotors, piston ring grooves, gears, gearbox bearings, connecting rods, and shift forks. The extended opposition for lightweight and excessive-performance substances became probably to increase the need for composites of the aluminum matrix [4].

SiC is one of the most useful and low-value reinforcements available for producing aluminum composites. The fact that SiC is highly extra-strong inside the aluminum matrix makes a higher choice for Al/SiC [5]. The low density, excessive energy, excessive modulus, low thermal enlargement, high put on resistance, and proper damping capacity of SiC particulate strengthened aluminum composites, in addition to the viability of diverse price-powerful synthesis routes, and have fueled substantial interest in these substances [6]. Thus far, researchers have investigated the conse-quences of SiC reinforcement in the micron-length period scale in the aluminum matrix composites and its commercial-grade alloys reporting success in improving the mechanical and thermal properties.

In the earlier studies on metal matrix composite has now moved forward toward the fabrication of the hybrid composite materials to fulfill the need of the manufac-turing companies to meet of cutting-edge technology like excellent electricity con-ductivity, less weight ratio, superior hardness, low density to weight ratio, and high wear resistance property [7]. Aluminum MMCs are proven to be a suitable candidate material for any manufacturers where the properties of the aluminum alloys had satisfied the manufacturer's requirements. They may be desired to have superior heat conduction, excellent strength, the acceptable range of damping factor values, and decrease density, which made these alloys as a best-fit material for the preparation of the various mechanical and other engineering components like cylinder block lining component, automobile force shafts, automotive piston parts, vehicle struc-tures, etc. Composites have come to be a sensible applicant as engineering regions for the market products, like automobiles; now they include the MMC components [8]. Aluminum matrix composites are considered to be used in vehicle engines because of advanced particular property and stiffness at progressed temperatures provided via the ceramic debris and particular energy by the Al-primarily based MMCs [9]. Processing of aluminum alloys extra than 2wt % of Si causes spheroidization of the crystallized eutectic alloy Si; consequently it changed into established that the addition of those foreign particles specifically ceramic particles offer advanced alloy property [10].

The dentric structure formation at the earlier stage inside the composite is noticed in Al-SiC composite, but location occurs in a quarter rich in particles; this damage causes particle breakup with interfacial decohesion [11]. The secondary processing of deformation results in the breakage of the particle (or whisker) and reduces the formation of the agglomerates. This action may produce more homogeneous particu-late distribution and enhance the bonding strength; overall these characteristics were found to have the improved mechanical property of these samples or the composite [12]. The introduction of SiC debris into Al alloys, however, effects an extensive reduction of their ductility and fracture toughness [13]. It has been widely popular that particle-strengthened MMC fails with the useful resource of matrix – reinforc-ing decohesion, reinforcement cracking, void forming, and coalescence ensuing in ductile matrix failure [14]. These stresses may additionally purpose plastic deforma-tion of the matrix or interfacial debonding of the matrix reinforcement resulting in early fracture and failure of these materials [15].

Poovazhagan et al., [16] studies concluded that addition SiC in Al matrix increase the wettability however, increment of SiC as reinforcement in the matrix improve the matrix hardness. Further, Mg becomes necessary to be added in a small quantity in the stir casting to have improved wettability mixtures. Including Mg as an alloying agent before applying SiC to the Al matrix will reduce the surface anxiety of molten aluminum and improve the SiC's wettability with softening aluminum [16]. Nie et al. described an examination of SiC nanoparticles that bolstered the composite magne-sium matrix. The microcrack may additionally provoke dense zones for the duration of the traction inside the SiC nanoparticle and it progresses to have low tensile observations at room temperature, ensuing in decreasing the elongation [17]. Afshin et al. studied the mechanical properties of aluminum and natural stable nano-composites for the microstructural arrangements and the performance on the room temperature. It has been stated that the SiC nanoparticles can drastically reduce the grain duration of the aluminum matrix at the proposed operating situation [18]. Brofin et al. have studied the addition of silicon to Mg-Al alloys to enhance the resistance to creep. It turned into believing the Mg_2Si particles are being usual through the AZ21 alloy solidification system [19].

Szaraz et al. expected 13 vol. % of silicon carbide particulates in the microstructure of magnesium alloy. The reinforcing process thus contributes to strengthening specifically via an extended density of dislocation[20]. The creeping activity of the infiltrated material was investigated by Pillai et al. The actual load can be calculated from the creeping mechanism, depending on the reinforcement content and the load implemented, resulting in the formation of a uniformly distributed particle inside the matrix with very low agglomeration [21].

Unsupervised learning in the form of observation is known as clustering. This clustering algorithm is being formed the different types of clusters by evaluating the rules. The data points within the clusters are similar and the data points in other clusters are dissimilar. The main objective of this clustering is to group similar data points together. The way of forming the clusters creates the different types of cluster techniques [22]. The types are partitioning methods and hierarchical methods.

8.1.1 Partitioning Cluster

Partitioning cluster method clusters the data points by calculating their distances from either n-dimensional plane or randomly or some particular points [23]. This method is used to form the optimal clusters from a large set of data as shown in Figure 8.1.

Examples of partition-based clustering methods include K-means, K-Medoids, CLARANS, etc.

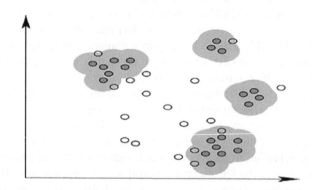

FIGURE 8.1   Partition-based clustering.

Global motion estimation on K-means clustering algorithms to improve the robustness features between the frames, to eliminate the redundancy data points. This makes the richness on the cluster and improves the computational efficiency [24]. Deterministic K-means (DK-means) initialization algorithm proposed to improve the results of classifications in terms of stability, faster and quality clusters when compared with MinMax and K-means ++ algorithms [25]. The Nystrom method with kernel K-means clustering algorithm identifies the dataset from different sources in a structured way than the linear clustering algorithm. This method produces computational clustering at a cheaper cost and maps the features in the high dimensional way [26].

Improved K-means clustering algorithm is to analyze the data in quick access compared with the existing algorithms. This method will evaluate the results based on the cluster range of information [27]. Centroid-based K-means cluster algorithms are used improve the analysis of finding accuracy and efficiency over the continuous dataset. Python programming language used the open-source library SciPy to improve the accuracy, stability, and performance of machine learning clustering algorithms [28]. Distance matrices (Euclidean distance or Manhattan distance) have been computing with the help of Python programming language over k-medoids clustering. It also computes the distance between the strings of characters, and to cluster only the information [29]. Python module Scikit is the one which helps to integrate the machine learning algorithms with the problems as supervised and unsupervised learnings. This package is on bringing the machine learning into ease of use, and speed up the performance, API consistency, and easier documentation [30].

The fast cluster package in Python library for machine learning clustering algorithms provides a faster implementation to make it more efficient than the existing packages as R Programming, Mathematica, and MATLAB. It saves the memory routines for K-means and hierarchical clustering data and also it improves the time complexity. The functionality part designed in replacement methods flashClust and hclust in R effort is less than the scipy.cluster.linkage in Python language [31]. The common challenging issues in machine learning is handling the class distributions in a balanced way. There are so many strategies to handle this problem with artificial data than the original data. The techniques which are used to modify the already trained data with the imbalanced classification of data is called a Synthetic Minority Oversampling Technique (SMOTE). This technique is used to avoid the outlier as well as noise data and effectively balance the classes. This method was implemented with the help of Python programming language [32]. Due to the enormous growth of the data which is taken from the outside world, lots of issues are created like difficult to manipulate and search the data. K-means algorithm is the popular clustering algorithm to improve the efficiency in massive datasets. Local representation of the K-means algorithm contains characteristics such as cardinality and center of mass on weighted data [33].

In the current chapter, the hard composite materials prepared on LM13 alloy with 63% (maximum solubility) of SiC were successfully fabricated through the ultrasonic-assisted stir casting process. The machinability of the composite and the surface topography on the AWJM machined composite surface was examined. To predict the suitable machining condition, the first time the machine learning language was used to reduce the mathematical complexity that arose at the time of optimizing the mechanical machining conditions. Python programming language code was used to sort the problems and to identify the suitable machining conditions for the fabricated composite. The observations are taken as per the Taguchi L27 orthogonal array with the varied parametric condition in abrasive water jet machining (AWJM), where here each AWJM parameter that was considered to have three independent levels. Hence 27 experimental runs were conditioned at the ambient condition using Garnet of particle size 80 as abrasive to cut the fabricated composite.

8.2   MATERIAL PREPARATION

Commercially available LM-13 (AlSi 7%) alloy was procured from the market in a rod form. 99.99% purity SiC was procured from the market that has a particle size of 15–30 micron. AlSi 7% was melted in an induction furnace at the temperature of 725°C. Before the introduction of silicon carbide reinforcement, the temperature of the furnace was reduced to 600°C. At a constant stirring condition of 200 RPM, the SiC was slowly added into the vortex of the mixture. The maximum solubility of 63wt. % SiC is added at a constant stirring condition. Before adding SiC into the matrix, SiC was preheated to a temperature of 900°C to remove the oxygen content and it was continuously maintained at 900°C for ten minutes and then it was introduced into the vortex. The entire manufacturing process was conducted in an argon atmosphere. After complete addition of 63% of SiC into the mixture, ultrasonic assistance is provided at a frequency rate of 2KHz. The vibration produced by the ultrasonicator enhanced the uniform deposition of the reinforcement in the mixture and avoid agglomeration. Permanent steel dies of 8 mm wall thick were preheated to a temperature of 100°C to improve the wettability of the molten mixture. Finally, the composite slurry was poured into the preheated die to obtain a composite ingot of dimension 25 cm × 25 cm × 1.2 cm even after the removal of the feeder head.

8.3   EXPERIMENTAL DESIGN AND OBSERVATIONS

AWJM has proven as a suitable machining process to cut any material irrespective of its properties. Machining operations such as producing holes, through the cut, blind holes, etc can be easily processed through AWJM. When compared to the plasma cutting process, AWJM process merits such as energy efficiency, low heat-affected zone, ease of machining process, work environmental safety, etc. The heat eliminated during plasma cutting operation is directly exposed to the open environment whereas a very low temperature would be experienced by AWJM while performing the cutting operation. AWJM has a large tank that was used to dilute the water pressure that exists from the nozzle. The residues are collected in the tank itself and it can be easily disposed of without affecting the ecosystem. Because of the aforesaid reason AWJM was used to call a green technology manufacturing process. However AWJM the thickness of the sample is a constraint to do the cutting process. An increase in the thickness of the sample creates a large taper cut surface and this too limited to some thickness. Higher the thickness of the sample lowers the machining performance. In the plasma cutting process, the thickness of the sample is not a constraint like that of AWJM.

AWJM becomes validated as an effective method because of its flexibility in cutting materials no matter the properties, removal of thermal effects at some stage in the technique, and minimal stresses that it imposes. Because it was a cold operating process, Abrasive Water Jet Machine (AWJM) was extensively favored to no heat-affected zones. AWJT serves as a replacement for the traditional cutter head of a turning check apparatus. The recent research on hybrid metal matrix composites has improved the properties of the material to a greater extent and finds its application in various engineering and the medical fields. The complexity in AWJM was getting the poor surface profile roughness value. AlSi7+63% SiC Composite turned into used as an experimentation material. This composite became prepared using dispersing silicon carbide and SiC in weight fraction in a liquid Al6061 aluminum alloy (matrix) through using the stir casting. A hybrid composite plate of length $25\,cm \times 25\,cm \times 1.2\,cm$ is cast evaluation of this material is done in X-ray radiography to verify that the materials were best and has no imperfection. In abrasive water jet device ingrained with a twin intensifier and pumping system trail experiments had been performed. The Garnet that has a particle size of 80 mesh is used as the cutting particle in the present study. AWJM has a tungsten carbide nozzle of 0.67 mm diameter with a length of 72 mm and it has an orifice diameter of 0.25 mm was used to enhance the pressure of the water. The entire unit was controlled by PLC. The nozzle and the sample are normal to each other for all the experimental runs. The experimental arrangements to cut AlSi7/SiC hybrid composite were shown in Figure 8.2.

The AWJM has various parameters involved in the determination of the machining performance. Among them, the most significant parameters that have effective

FIGURE 8.2   Abrasive water jet machine.

involvement in the prediction of the output responses alone were to be considered as the machining parameter in the present work and this was computed based on the literature survey. The following parameters are taken into consideration.

Water Pressure (WP)
Cutting Distance (CD)
Cutting Speed (CS)

The output responses are:

Material Removal Rate (MRR)
> The MRR was the actual quantity of the materials that are removed during the machining process. Usually, in the AWJM, the materials are eroded along the boundary of the water beam. Usually, the removal of materials at AWJM will be in particle form.

Kerf Width Angle (KWA)
> The KWA was the angle formed during the high-pressure water beam follow at the time of cut. In AWJM KWA was one of the drawbacks and it has to be optimized by the selection of the proper machining conditions. Usually, it was calculated based on the measurement of the width of cut both the measurement taken from the top surface and the bottom surface of the cut sample. Higher the thickness of the sample and hardness property predominantly determines the KWA.

Surface Roughness (Ra)
> Generally in AWJM, the newly formed surface as the result of machining was divided into three zones.

> The preliminary deformation region (PDR): It was the region where the abrasive particle initially starts cutting the sample. The continuous forging effect of the Garnet particles particularly reduces the surface profile roughness of the sample in this region. The excess flow of backscattered abrasives is witnessed in these regions progress to have excess MRR compared to the other region. Parallelly it will increase the KWA of the machined surface. Increase in the thickness of the cut samples cause to have excess flow of backscattered water. As excess back flow water is a mix of abrasives that result in wider cut.

> Easy Cutting Region (ECR), is the center portion of the machine surface. For example, if the sample thickness was 10 mm means the ECR zone will start from 4 mm from the top cut surface and it extends up to 8 mm. This region will face the energy of the abrasive, as an increase in depth of cut the partial energy loss abrasives will produce some machining effect at these regions. However, these surfaces will have an acceptable range of surface finish compared to the other two regions. Each abrasive removed a small quantity of material at low occurrence angles, and hence, the surface finish in these regions is significantly better than the rough-cut surrounding area. This could result in a good surface finish at this place.

Tough Reduce Region (TCR)

This was the region of the machined surface, and it indicated the bottom-most portion of the cut. The surface defects in these regions are very high-profile roughness values. The presence of an excess amount of scar mark, wear track, and jet curves are visible in these regions. Typically, the collision of particles at high angles alternates the mode of material elimination. Wear and tear are produced through a deformation mechanism to cause striation and rough slicing marks that could be created along the diverging water line. From determine three, observers can recognize the results of abrasives produced in one of a kind kerf areas. The growth in kerf taper was determined with growth within the thickness of the cut samples.

Figure 8.3 show the AWJM cut surface. IDR caused by piercing and impingement of high pressure abrasive water flow on this region. Smooth cutting region (SCR) is formed due to the jet, which may produce a very fine surface and striated and coarse features cause by the high nozzle movement in RCR, the arrow indicates the cutting direction.

8.3.1 SURFACE ROUGHNESS TESTER

The surface profile was estimated using surface profile analyzer size of Mitutoyo made SJ-411 which has a scope of 350 μm with a test pace of 0.25 mm/s over a range of 5 mm. Three perceptions are made at PDR, ECR, and TCR of the cut area, and the midpoints of those are organized in Ra. The use of the profile projector, the shape of the cut surface was determined. Figure 8.4 shows the surface profile roughness analyzer.

AUX 220 of Shimadzu made with high accuracy rate of 0.0001 micro meters was used to measure the degree the composite machined surface.

Cutting Speed (CS), water pressure (WP), and cutting distance (CD) are picked boundaries with three degrees of each in AWJM. The picked boundaries are recorded in Table.8.1.

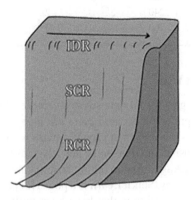

FIGURE 8.3 Surface roughness profile of AWJM.

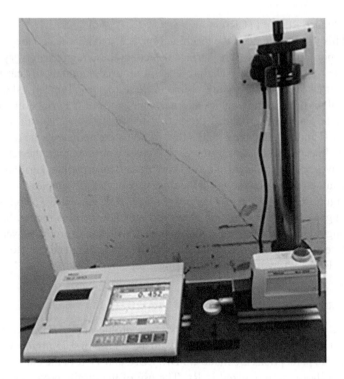

FIGURE 8.4   Surface roughness tester.

TABLE 8.1
AWJM Machining Parameters and the Ranges

AWJM Parameters

Levels	WP (bar)	CD (mm)	CS (mm/s)
1	220	1	20
2	240	2	30
3	260	3	40

8.4   PYTHON ANALYSIS

There are so many strategies for all data science ventures, but the Python programming language is more preferred because there are many advantages over all other languages. The latest version of Python provides users with more than 70,000 packages available on the website, and may make new changes to any of the applications. Some of the important packages used in our manuscript are Scikit Learn, Matplotlib, Pandas, NumPy, and NdArray. Scikit Learn is a powerful package that contains most of the Machine Learning algorithms that have been built in. Matplotlib is a dram package used for all types of plotting to implement the output model. The Pandas package supported the results of the data analysis. Mostly, all pre-processing operations are performed by Pandas. NumPy, which provides the Nd Array string, is one of the most popular Python packages. It offers a range of all forms of service. In Python, the user needs to focus on using modules instead of implementing algorithms or models. It reduces almost 50% of the user's implementation time. Python is best suited for machine learning because of all the above. Python is easy to maintain and easy to use.

From the machining Table 8.2, we used three parameters water pressure (WP), cutting distance (CD), and cutting speed (CP) to determine the output labels material removal rate (MRR), kerf angle (Ka), and surface profile roughness (Ra). With three inputs and nine combinations, experiments were conducted to messure the 3 output responses. To implement the combinations we used Python programming technology. It is easy to work with two attributes (2-D) instead of three attributes to draw better conclusions. Every time two combinations of input variables are used, like (WP, CD), (CD, CP), (WP, CP) vice versa. To conclude every time it is not possible to get the results concerning three output variables, so we used to conclude with the respective two output variables only. Scikit learn and SciPy packages are used to implement agglomerative clustering and to draw the dendrogram. Scikit learn is a standard AI bundle; it gives information on AI practices like model fitting, forecasting, cross-approval. SciPy is a Python library utilized for logical and specialized registering. SciPy contains modules for improvement, direct variable based math, mix, introduction, exceptional capacities, FFT, sign and picture preparing, ODE solvers, and different undertakings normal in science and building. In model agglomerative bunching we utilized several groups or three and utilized Euclidean separation measure to make the individual groups into a remarkable bunch. Each time we discover ward separation in information network to make groups.

To represent the clusters phenomena we used to draw dendrogram with the ward distance used to make the clusters. Instead of measuring the distance directly; it analyzes the variance of clusters. For quantitative variables, Ward's method is said to be the most suitable method. K-means another method we are used to making the clusters. In this we used $k = 3$ means we are making the initial three clusters. For clustering, the first combination is CD, and CS, here total number of samples are 10 with standard deviation 0.02. Initial we clusters $c1 = (1, 20)$, $c2 = (2, 30)$, and $c3 = (3, 40)$ after calculation of Euclidean distance. Here we got minimum distance data point into the cluster.

Experimental results on Analyzing AHC using Python

```
WP and CD as input
import pandas as p
import numpy as n
from matplotlib import pyplot as pl
from sklearn.cluster import AgglomerativeClustering
import scipy.cluster.hierarchy as sch
data_set = p.read_csv(r'C:\Users\Experimentalvalues\Desktop\
data.csv')
V = data_set.iloc[:, [0, 1]].values
```

```
dendrogram = sch.dendrogram(sch.linkage(V, method='ward'))
model = AgglomerativeClustering(n_clusters=3,
affinity='euclidean', linkage='ward')
model.fit(V)
labels = model.labels_labels
array([2, 2, 2, 2, 2, 2, 2, 2, 2, 1, 1, 1, 1, 1, 1, 1, 1, 1,
0, 0, 0, 0, 0, 0, 0, 0, 0], dtype=int64)
pl.scatter(V[labels==0, 0], V[labels==0, 1], s=50,
marker='o', color='red')
pl.scatter(V[labels==1, 0], V[labels==1, 1], s=50,
marker='o', color='blue')
pl.scatter(V[labels==2, 0], V[labels==2, 1], s=50,
marker='o', color='green')
pl.scatter(V[labels==3, 0], V[labels==3, 1], s=50,
marker='o', color='purple')
pl.scatter(V[labels==4, 0], V[labels==4, 1], s=50,
marker='o', color='orange')
pl.show()
```

Considering CD and CS as input parameters will provide 3 classes as clustering output. The response variables are clustered into 3 classes. They are first nine observations 0–8 are under class 1, observations 9–17 are under class 2, and observations 18–26 are under class 3 as shown in Figure 8.5.

```
CD and CS as inputs
import pandas as p
import numpy as n
from matplotlib import pyplot as pl
from sklearn.cluster import AgglomerativeClustering
```

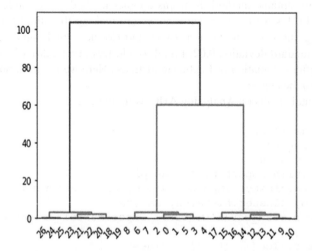

FIGURE 8.5   Dendrogram for WP and CD.

```
import scipy.cluster.hierarchy as sch
V = data_set.iloc[:, [1, 2]].values
dendrogram = sch.dendrogram(sch.linkage(V, method='ward'))
model = AgglomerativeClustering(n_clusters=3,
affinity='euclidean', linkage='ward')
model.fit(V)
labels = model.labels_labels
array([2, 1, 0, 2, 1, 0, 2, 1, 0, 2, 1, 0, 2, 1, 0, 2, 1, 0,
2, 1, 0, 2,1, 0, 2, 1, 0], dtype=int64)
pl.scatter(V[labels==0, 0], V[labels==0, 1], s=50,
marker='o', color='red')
pl.scatter(V[labels==1, 0], V[labels==1, 1], s=50,
marker='o', color='blue')
pl.scatter(V[labels==2, 0], V[labels==2, 1], s=50,
marker='o', color='green')
pl.scatter(V[labels==3, 0], V[labels==3, 1], s=50,
marker='o', color='purple')
pl.scatter(V[labels==4, 0], V[labels==4, 1], s=50,
marker='o', color='orange')
pl.show()
```

Considering CD and CS as input parameters will provide three classes as clustering output. The response variables are partitioned into three classes that are first observation 0 under class 1, observation 1 under class 2, observation 2 under class 3, and observation 3 under class 1, the given 27 observations are classified and it is shown in Figure 8.6.

```
WP and CS as inputs
V = data_set.iloc[:, [0, 2]].values
dendrogram = sch.dendrogram(sch.linkage(X, method='ward'))
model = AgglomerativeClustering(n_clusters=3,
affinity='euclidean', linkage='ward')
model.fit(V)
```

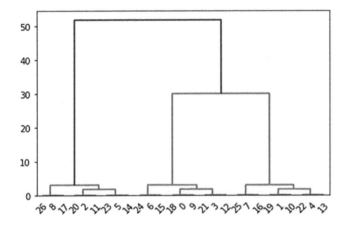

FIGURE 8.6   Dendrogram for CD and CS.

```
labels = model.labels_labels
array([2, 2, 2, 2, 2, 2, 2, 2, 2, 1, 1, 1, 1, 1, 1, 1, 1, 1,
0, 0, 0, 0, 0, 0, 0, 0, 0], dtype=int64)
pl.scatter(V[labels==0, 0], V[labels==0, 1], s=50,
marker='o', color='red')
pl.scatter(V[labels==1, 0], V[labels==1, 1], s=50,
marker='o', color='blue')
pl.scatter(V[labels==2, 0], V[labels==2, 1], s=50,
marker='o', color='green')
pl.scatter(V[labels==3, 0], V[labels==3, 1], s=50,
marker='o', color='purple')
pl.scatter(V[labels==4, 0], V[labels==4, 1], s=50,
marker='o', color='orange')
pl.show()
```

Taking WP and CS as input parameters and produces the three classes as clustering output. The response variables are partitioned into three classes that are first nine observations 0–8 are under class 1, observations 9–17 are under class 2, and observations 18–26 are under class 3 as shown in Figure 8.7.

Experimental results on analyzing K-means algorithm using Python:

Partition-based clustering method (K-means) is a very popular and simple algorithm.

```
CD and CS as input
import pandas as p
import numpy as n
from sklearn.cluster import KMeans
from sklearn.preprocessing import LabelEncoder
from sklearn.preprocessing import MinMaxScaler
import seaborn as sns
import matplotlib.pyplot as plt
```

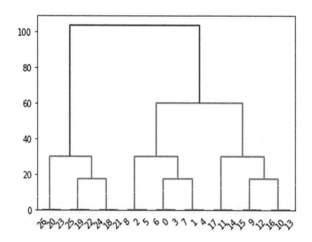

FIGURE 8.7   Dendrogram for WP and CS.

```
%matplotlib inline from sklearn.datasets.samples_generator
import make_blobs
X, _ = make_blobs(n_samples=10, centers=3, n_features=2,
cluster_std=0.2, random_state=0)
objnames = []
for i in range(1, 28):
obj = "Object " + str(i)
objnames.append(obj)
data = p.DataFrame({
'Object': objnames,
'CD': [1,1,1,2,2,2,3,3,3,1,1,1,2,2,2,3,3,3,1,1,1,2,2,2,3,3,],
'CS':
[20,30,40,20,30,40,20,30,40,20,30,40,20,30,40,20,30,40,20,30,4
0,20,30,40,20,30,40]})
c0 = (1, 20)
c1 = (2, 30)
c2 = (3,40)
def calculatedistance(centroid, X, Y):
distances = []
cx, cy = centroid
for x, y in list(zip(X, Y)):
root_diff_x = (x - cx) ** 2
root_diff_y = (y - cy) ** 2
distance = n.sqrt(root_diff_x + root_diff_y)
distances.append(distance)
return distances
data['MRR'] = calculatedistance(c0, data.CD, data.CS)
data['KA'] = calculatedistance(c1, data.CD, data.CS)
data['Ra'] = calculatedistance(c2, data.CD, data.CS)
data['Cluster'] = data[['MRR', 'KA', 'Ra']].apply(n.argmin,
axis =1)
data['Cluster'] = data['Cluster'].map({'MRR': 'C1', 'KA':
'C2', 'Ra': 'C3'})
x_newcentroid1 = data[data['Cluster']=='C1']['CD'].mean()
y_newcentroid1 = data[data['Cluster']=='C1']['CS'].mean()
x_newcentroid2 = data[data['Cluster']=='C3']['CD'].mean()
y_newcentroid2 = data[data['Cluster']=='C3']['CS'].mean()
print('Centroid 1 ({}, {})'.format(x_newcentroid1,
y_newcentroid1))
print('Centroid 2 ({}, {})'.format(x_newcentroid2,
y_newcentroid2))
output :
Centroid 1 (2.0, 20.0)
Centroid 2 (2.0, 40.0)
```

Considering CD and CS as input parameters will get 2 centroids as response values in which Centroid 1 contains 2.0, 20.0 and Centroid 2 contains 2.0, 40.0 as the best value.

```
WP and CD as input
import pandas as pd
import numpy as n
```

```
from sklearn.cluster import KMeans
from sklearn.preprocessing import LabelEncoder
from sklearn.preprocessing import MinMaxScaler
import seaborn as sns
import matplotlib.pyplot as plt
%matplotlib inline
from sklearn.datasets.samples_generator import make_blobs
X, _ = make_blobs(n_samples=10, centers=3, n_features=2,
cluster_std=0.2, random_state=0)
objnames = []
for i in range(1, 28):
obj = "Object " + str(i)
objnames.append(obj)
c0 = (220, 1)
c1 = (240, 2)
c2 = (260, 3)
def calculatedistance(centroid, X, Y):
distances = []
cx, cy = centroid
for x, y in list(zip(X, Y)):
root_diffx = (x - cx) ** 2
root_diffy = (y - cy) ** 2
distance = n.sqrt(root_diffx + root_diffy)
distances.append(distance)
return distances
data['MRR'] = calculatedistance(c0, data.WP, data.CD)
data['KA'] = calculatedistance(c1, data.WP, data.CD)
data['Ra'] = calculatedistance(c2, data.WP, data.CD)
data['Cluster'] = data[['MRR', 'KA', 'Ra']].apply(n.argmin,
axis =1)
data['Cluster'] = data['Cluster'].map({'MRR': 'C1', 'KA':
'C2', 'Ra': 'C3'})
x_newcentroid1 = data[data['Cluster']=='C1']['WP'].mean()
y_newcentroid1 = data[data['Cluster']=='C1']['CD'].mean()
x_newcentroid2 = data[data['Cluster']=='C3']['WP'].mean()
y_newcentroid2 = data[data['Cluster']=='C3']['CD'].mean()
print('Centroid 1 ({}, {})'.format(x_newcentroid1,
y_newcentroid1))
print('Centroid 2 ({}, {})'.format(x_newcentroid2,
y_newcentroid2))
Output:
Centroid 1 (220.0, 2.0)
Centroid 2 (260.0, 2.0)
```

Considering WP and CD as input parameters will get 2 centroids as response values in which Centroid 1 contains 220.0, 2.0 and Centroid 2 contains 260.0, 2.0 as the best value.

```
WP and CS as input
import pandas as pd
import numpy as n
```

```
from sklearn.cluster import KMeans
from sklearn.preprocessing import LabelEncoder
from sklearn.preprocessing import MinMaxScaler
import seaborn as sns
import matplotlib.pyplot as plt
%matplotlib inline from sklearn.datasets.samples_generator
import make_blobs
X, _ = make_blobs(n_samples=10, centers=3, n_features=2,
cluster_std=0.2, random_state=0)
objnames = []
for i in range(1, 28):
obj = "Object " + str(i)
objnames.append(obj)
c0 = (220, 20)
c1 = (240, 30)
c2 = (260, 40)
def calculatedistance(centroid, X, Y):
distances = []
cx, cy = centroid
for x, y in list(zip(X, Y)):
root_diffx = (x - cx) ** 2
root_diffy = (y - cy) ** 2
distance = n.sqrt(root_diffx + root_diffy)
distances.append(distance)
return distances
data['MRR'] = calculatedistance(c0, data.WP, data.CS)
data['KA'] = calculatedistance(c1, data.WP, data.CS)
data['Ra'] = calculatedistance(c2, data.WP, data.CS)
data['Cluster'] = data[['MRR', 'KA', 'Ra']].apply(n.argmin,
axis =1)
data['Cluster'] = data['Cluster'].map({'MRR': 'C1', 'KA':
'C2', 'Ra': 'C3'})
x_newcentroid1 = data[data['Cluster']=='C1']['WP'].mean()
y_newcentroid1 = data[data['Cluster']=='C1']['CS'].mean()
x_newcentroid2 = data[data['Cluster']=='C3']['WP'].mean()
y_newcentroid2 = data[data['Cluster']=='C3']['CS'].mean()
print('Centroid 1 ({}, {})'.format(x_newcentroid1,
y_newcentroid1))
print('Centroid 2 ({}, {})'.format(x_newcentroid2,
y_newcentroid2))
OUTPUT
Centroid 1 (220.0, 30.0)
Centroid 2 (260.0, 30.0)
```

Considering WP and CS as input parameters will get 2 centroids as response values in which Centroid 1 contains 220.0, 30.0 and Centroid 2 contains 260.0, 30.0 as the best value.

A disperse plot of considered experimental conditions is shown in Figure 8.8. Here uncovers connections or relationships between two factors. Such connections show themselves by any non-irregular structure in the plot. In this to plot the information we utilized a scotter plot. Every time we used two input combinations every

Object	WP (bar)	CD (mm)	CS (mm/min)	CD and CS as input				CD and WP as input				WP and CS as input			
				MRR (g/s)	KA (Deg)	Ra (μm)	Class	MRR (g/s)	KA (Deg)	Ra (μm)	Class	MRR (g/s)	KA (Deg)	Ra (μm)	Class
1	220	1	20	0.000	10.049	20.09	C1	0.00	20.02	40.1	C1	0.0	22.36	44.72	C1
2	220	1	30	10.00	1.000	10.19	C2	0.00	20.02	40.1	C1	10.0	20.0	41.23	C1
3	220	1	40	20.00	10.049	2.00	C3	0.00	20.02	40.1	C1	20.0	22.36	40.0	C1
4	220	2	20	1.000	10.000	20.02	C1	1.00	20.00	40.0	C1	0.0	22.36	44.72	C1
5	220	2	30	10.049	0.000	10.05	C2	1.00	20.00	40.0	C1	10.0	20.0	41.23	C1
6	220	2	40	20.05	10.00	1.00	C3	1.00	20.00	40.0	C1	20.0	22.36	40.0	C1
7	220	3	20	2.00	10.05	20.00	C1	2.00	20.02	40.0	C1	0.0	22.36	44.72	C1
8	220	3	30	10.19	1.00	10.00	C2	2.00	20.02	40.0	C1	10.0	20.0	41.23	C1
9	220	3	40	20.09	10.05	0.00	C3	2.00	20.02	40.0	C1	20.0	22.36	40.0	C1
10	240	1	20	0.000	10.049	20.09	C1	20.0	1.00	20.1	C2	20.0	10.0	28.28	C2
11	240	1	30	10.00	1.000	10.19	C2	20.0	1.00	20.1	C2	22.36	0.00	22.36	C2
12	240	1	40	20.00	10.049	2.00	C3	20.0	1.00	20.1	C2	28.28	10.0	20.0	C2
13	240	2	20	1.000	10.000	20.02	C1	20.0	0.00	20.0	C2	20.0	10.0	28.28	C2
14	240	2	30	10.049	0.000	10.05	C2	20.0	0.00	20.0	C2	22.36	0.00	22.36	C2
15	240	2	40	20.05	10.00	1.00	C3	20.0	0.00	20.0	C2	28.28	10.0	20.0	C2
16	240	3	20	2.00	10.05	20.00	C1	20.1	1.00	20.0	C2	20.0	10.0	28.28	C2
17	240	3	30	10.19	1.00	10.00	C2	20.1	1.00	20.0	C2	22.36	0.00	22.36	C2
18	240	3	40	20.09	10.05	0.00	C3	20.1	1.00	20.0	C2	28.28	10.0	20.0	C2
19	260	1	20	0.000	10.049	20.09	C1	40.0	20.02	2.00	C3	40.0	22.36	20.0	C3
20	260	1	30	10.00	1.000	10.19	C2	40.0	20.02	2.00	C3	41.23	20.0	10.0	C3
21	260	1	40	20.00	10.049	2.00	C3	40.0	20.02	2.00	C3	44.72	22.36	0.00	C3
22	260	2	20	1.000	10.000	20.02	C1	40.0	20.00	1.00	C3	40.0	22.36	20.0	C3
23	260	2	30	10.049	0.000	10.05	C2	40.0	20.00	1.00	C3	41.23	20.0	10.0	C3
24	260	2	40	20.05	10.00	1.00	C3	40.0	20.00	1.00	C3	44.72	22.36	0.00	C3
25	260	3	20	2.00	10.05	20.00	C1	40.1	20.02	0.00	C3	40.0	22.36	20.0	C3
26	260	3	30	10.19	1.00	10.00	C2	40.1	20.02	0.00	C3	41.23	20.0	10.0	C3
27	260	3	40	20.09	10.05	0.00	C3	40.1	20.02	0.00	C3	44.72	22.36	0.00	C3

FIGURE 8.8   Disperse plot for the 27 experimental runs.

time to plot the data like (WP, CD), (CD, CS), and (WP, CS). If the data plotted in normal distribution then it will produce good results. If data correctly plotted or not here we used scotter plot. If the distribution of area of data is very less then it is good for any model. The scotter plot data used to draw invariant data as well as binomial data. It can be imported from the matplot library in Python.

8.5   RESULTS AND DISCUSSIONS

Proposed K-means and agglomerative clustering machine learning algorithms on machining parameters through Python programming language to improve the accuracy of the response values. It has been absolved that the classes of clustering algorithms produce the best combinational input values to produce improved results. Python programming languages on machining parameters to understand the consistency of the response variables and evaluate the performance of the machining parameters. Python in machine learning algorithms produces fast clusters when compared with existing packages such as R programming, MATLAB, and Mathematica.

Step-by-step K-means algorithm:

Step 1: Initialize the centroids randomly as shown in Figure 8.9. Let's assume k=3, and name three centroid points for three clusters (C1, C2, C3).

Initialize the centroids

```
C1 = (-1, 4)
C2 = (-0.2, 1.5)
C3 = (2, 2.5)
```

TABLE 8.2

Experimental Observations

Ex. Nos.	WP	CD	CS	MRR	KWA	Ra
1	220	1	20	0.0042	0.2376	3.447
2	220	1	30	0.0055	0.1623	3.457
3	220	1	40	0.0070	0.2608	3.097
4	220	2	20	0.0047	0.3147	3.246
5	220	2	30	0.0060	0.2739	3.346
6	220	2	40	0.0083	0.3743	3.525
7	220	3	20	0.0048	0.5602	3.595
8	220	3	30	0.0067	0.6750	3.490
9	220	3	40	0.0081	0.5234	3.586
10	240	1	20	0.0046	0.5736	2.963
11	240	1	30	0.0055	0.2000	3.333
12	240	1	40	0.0084	0.2768	3.401
13	240	2	20	0.0046	0.5021	3.206
14	240	2	30	0.0064	0.3054	3.503
15	240	2	40	0.0086	0.9212	3.622
16	240	3	20	0.0045	0.7042	3.349
17	240	3	30	0.0072	0.7487	3.118
18	240	3	40	0.0099	0.7906	3.757
19	260	1	20	0.0044	0.5881	4.645
20	260	1	30	0.0059	0.5877	4.259
21	260	1	40	0.0078	0.4554	4.915
22	260	2	20	0.0043	0.5554	4.754
23	260	2	30	0.0065	0.3014	3.908
24	260	2	40	0.0083	0.5503	4.093
25	260	3	20	0.0052	0.7393	3.996
26	260	3	30	0.0068	0.3481	3.928
27	260	3	40	0.0090	0.5703	3.780

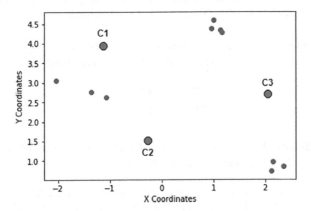

FIGURE 8.9   Random initialization of centroids.

Step 2: To figure the separation between the centroids and every information point either by utilizing Euclidean or Manhattan separation. Python code for computing Euclidean separation between the information focuses.

A helper function to calculate the Euclidean distance between the data points and the centroids

```
def calculate_distance(centroid, X, Y):
distances = []
# Unpack the x and y coordinates of the centroid
c_x, c_y = centroid
Iterate over the data points and calculate the distance using
the given formula for x, y in list(zip(X, Y)):
root_diff_x = (x - c_x) ** 2
root_diff_y = (y - c_y) ** 2
distance = np.sqrt(root_diff_x + root_diff_y)
distances.append(distance)
return distances
# Calculate the distance and assign them to the DataFrame
accordingly
data['C1_Distance'] = calculate_distance(c1, data.X_value,
data.Y_value)
data['C2_Distance'] = calculate_distance(c2, data.X_value,
data.Y_value)
data['C3_Distance'] = calculate_distance(c3, data.X_value,
data.Y_value)
```

Step 3: Compare the distance to find the minimum distance of each data points to each centroids and assign into the respective centroids. Repeat these steps until all the data points make it as cluster.

\# Get the base separation centroids data['Cluster']=data[['C1_Distance', 'C2_Distance', 'C3_Distance']].apply(np.argmin, pivot =1)

\# Map the centroids as needs be and rename them data['Cluster']=data['Cluster'].map({'C1_Distance': 'C1', 'C2_Distance': 'C2', 'C3_Distance': 'C3'})

Step 4: if there is no data points belongs to any cluster, then update the centroids by calculating the mean of the coordinates by using Equation 1.

$$\frac{\sum_{i=1}^{n} x_i}{n}, \frac{\sum_{i=1}^{n} y_i}{n} \tag{8.1}$$

x_i and y_i are the coordinates and n is the number of data points in the cluster. Coding for calculating mean for all the three clusters

```
# Calculate the coordinates of the new centroid from cluster 1
x_new_centroid1 = data[data['Cluster']=='C1']['X_value'].
mean()
```

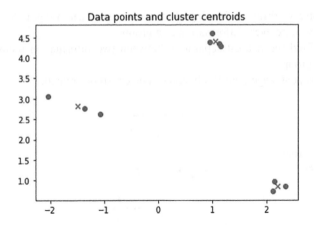

FIGURE 8.10 Plot between the cluster centroids and data points.

```
y_new_centroid1 = data[data['Cluster']=='C1']['Y_value'].
mean()
# Calculate the coordinates of the new centroid from cluster 2
x_new_centroid2 = data[data['Cluster']=='C2']['X_value'].
mean()
y_new_centroid2 = data[data['Cluster']=='C2']['Y_value'].
mean()
# Calculate the coordinates of the new centroid from cluster 3
x_new_centroid3 = data[data['Cluster']=='C3']['X_value'].
mean()
y_new_centroid3 = data[data['Cluster']=='C3']['Y_value'].
mean()
Print the coordinates of the new centroids
print('Centroid 1 ({}, {})'.format(x_new_centroid1,
y_new_centroid1))
print('Centroid 2 ({}, {})'.format(x_new_centroid2,
y_new_centroid2))
print('Centroid 3 ({}, {})'.format(x_new_centroid3,
y_new_centroid3))
```

Centroids in the K-means algorithm can be updated at every iteration based on the mean value. The data points can make it a cluster based on newly updated centroid points as shown in Figure 8.10.

8.6 AGGLOMERATIVE HIERARCHICAL CLUSTERING

Grouping of similar objects called hierarchical clusters and its two types namely agglomerative and divisive are shown in Figure 8.11. An agglomerative Hierarchical Cluster is a base-up approach (begins from numerous small leaves and consolidate to shape a major group). Disruptive Hierarchical Cluster is a top-down methodology (begins from a solitary enormous group and breaks into a modest number of bunches) [34].

Step-by-step working on forming of the cluster is shown in Figure 8.12.

Stage 1: Assume every information as a group.

Stage 2: Find the nearest separation between two information focuses and put them as one group.

Stage 3: Repeat stage 2 until it becomes one enormous bunch.

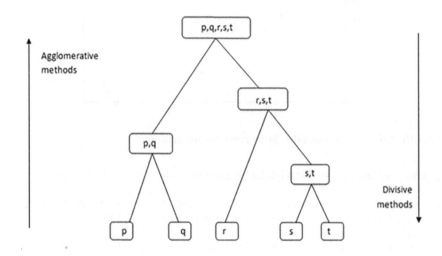

FIGURE 8.11   Agglomerative and divisive clustering.

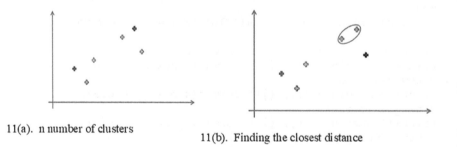

11(a). n number of clusters

11(b). Finding the closest distance

11(c). One large cluster

FIGURE 8.12   Forming of clusters.

The dendrogram is used to envisage the grouping of clusters as well as history and it builds the number of clusters optimally. First, it establishes the biggest vertical distance that should not traverse to any other clusters. Then the horizontal line has been drawn at both edges [35].

Several vertical lines on the horizontal lines = an optimal number of clusters.

The separation between the groups can be determined by utilizing linkage models [36]. It is four sorts to be specific, single linkage, normal linkage, total linkage, and ward linkage have appeared in Figure 8.13. The various sorts of results created by the four linkages appear in Figure 8.14.

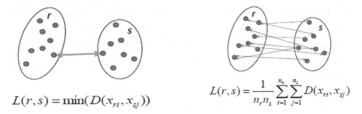

$$L(r,s) = \min(D(x_{ri}, x_{sj}))$$

8.12(a) Single Linkage

$$L(r,s) = \frac{1}{n_r n_s} \sum_{i=1}^{n_r} \sum_{j=1}^{n_s} D(x_{ri}, x_{sj})$$

8.12(b) Average Linkage

$$L(r,s) = \max(D(x_{ri}, x_{sj}))$$

8.12(c) Complete Linkage

8.12(d) Ward Linkage

FIGURE 8.13 Types of linkage criteria.

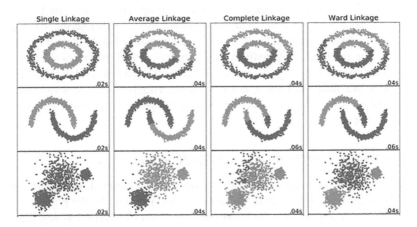

FIGURE 8.14 Different results created by different types of linkage.

The shortest distance between the two data points is used to calculate through either Euclidean distance or Manhattan distance. For example, if $a = (x, y)$ and $b = (p, r)$, the Euclidean distance between a and b is $\sqrt{(x-p)^2 + (y-r)^2}$ and the Manhattan distance between a and b is $|x-p| + |y-r|$.

Common clustering methods like K-means (partition-based method) and Agglomerative Hierarchical Clustering using Python are applied to machining parameters to analyze the several components on algorithms [37]. Using the Jupyter Notebook tool to execute the code of AHC and K-means algorithms in Python. Hierarchical clustering is an important clustering on computational science that used to store the quadratic space of matrix which depends on the number of objects. SparseHC clustering algorithm is proposed to scan the sorted data matrix step by step. The dendrogram is built by combining the pairs of the cluster that depends on the smallest distance between the data points. This method achieved in reducing the memory complexity to improve the efficiency of the datasets when compared with existing algorithms [38].

Most of the researches worked on various issues in machine learning algorithms. From the clustering algorithms, Agglomerative Hierarchical Clustering is the one to improve the performance in real-time applications. It is used to solve the problem of clustering in a realistic time [39]. Mapping manually in machine learning is very expensive, consumes more time and a huge gap is created between the functionality of the systems and implementation. So, Python language in machine learning clustering algorithms can automatically visualize and construct the graph for a given database; execution paths will be clustered for graph abstractions and clusters are labeled based on their functional behaviors. This work is used to build the gap between the system functionality and the graph to improve the feasibility of the clusters [40]. The research on information retrieval in web clustering is a major issue. Because the searching for results from the huge amount of data is very difficult. Search Results Clustering (SRC) with Agglomerative Hierarchical Clustering aims to retrieve the data from the clusters very faster. This will improve the quality of clusters and reduce the time complexity [41].

Agglomerative Hierarchical Clustering with average linkage is used for only cluster the data which is huge in number and not for scaling. The dendrogram is the best algorithm that reduces the complexity of quadratic which will improve the spaces of the matrix. Twister tries in dendrogram is very sensitive but it is possible in hash functions. The meta-heuristic methodology is proposed and integrates a fitness surrogate model to enhance the performance of computational time [42]. Real-time data is very huge in number and it is a lack of effective visualization. Meta-visualization Agglomerative Hierarchical Clustering uses dendrogram to represent the features without outlier [43].

8.7   SURFACE TOPOGRAPHY ON KERF SURFACE

Figure 8.15 shows the microstructural portrayal of the machined surface. In Figure 8.15a, the small scale level breaks which can be formed close by the grain limit are displayed. Barely any small scale breaks are seen over the kerf areas and they are the proof that the thermal effect is created span the hour of machining while at the

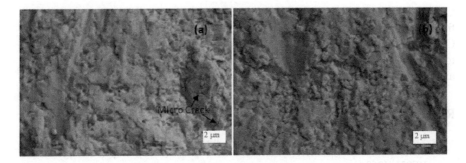

FIGURE 8.15   Sample AWJM cut surface.

same time utilizing Garnet as abrasives. The increase in the temperature and ensuing cooling produce thermal distortion. The high extended Garnet abrasives import its vitality to the composite surface on machining which comes out in pressing and squashing of composite particles. They cause the benefit of a fresh out of the new kerf regions. Figure 8.15b shows the intense diminished rough-cut areas at high levels of WP. The collusion of garnet inside the water jet before reach the cut region and makes jet disparity. Those low hardness abrasives, simultaneously as on machining the composite shade into little bits, bring about furrowing impact and imprint put on the tune at the diminished surface everywhere micron run. The abrasive wear component especially depends on the hardness of rough particles and the cut material. While the utilization of Garnet in a high running condition, the full energy got abrasives to sway over this area and it is decreased widely because of CS.

8.8   CONCLUSIONS

High-value forecasts can drive better decision-making and correct real-time action without human intervention. Machine learning as technology helps analyze vast amounts of data, eases the tasks of data scientists in an automated process, and gains a lot of popularity and recognition. Machine learning has changed the way data extraction and analysis work by involving an artificial collection of common methods that have replaced conventional statistical techniques.

Machine learning algorithms have been successfully used in various optimization processes, monitoring and control applications in manufacturing, and predictive maintenance in different industries. This technique was found to provide decision-making for improved optimization of machining parameters, particularly in the search for better optimal solutions for a small dataset. The advantage of using this machine learning technique on machining parameters is to increase the usability of applications in many cases, which is relatively easy to use and increases the classification performance by adjusting the parameters comfortably.

The mathematical complexity in optimization the machining condition was greatly eliminated using machine learning classifications K-means and AHC in a Python programming language. For the given set of problems associated with AWJM for L27 Orthogonal Array to AlSi7/63% SiC composite, we get final centroids are

C1 = (2.0, 20.0) and C2 = (2.0,40.0). Like that remaining combinations WP, CD the final centroids are C1 = (220.0,2.0) and C2 = (260.0,2.0) and the combination WP, CS the final centroids are C1 = (220.0, 30.0) and C2 = (260.0,30.0). Sample surface topography on the cut surface shows the excellent bonding occurred in the matrix and has a plastic deformation surface.

REFERENCES

1. Luo, A. Processing, microstructure, and mechanical behavior of cast magnesium metal matrix composites. *MMTA* 26, 2445–2455 (1995). https://doi.org/10.1007/BF02671259.
2. Ajith Kumar, K.K., Viswanath, A., Rajan, T.P.D. et al. Physical, mechanical, and tribological attributes of stir-cast AZ91/SiCp composite. *Acta Metallurgica Sinica (English Letters)* 27, 295–305 (2014). https://doi.org/10.1007/s40195-014-0045-3.
3. Ye, H., Liu, X.Y. Review of recent studies in magnesium matrix composites. *Journal of Materials Science* 39, 6153–6171 (2004). 10.1023/B:JMSC.0000043583.47148.31.
4. Chua, B. W., Lu, L., Lai, M. O. Influence of SiC particles on mechanical properties of Mg based composite. *Composite Structures* 47, 595–601 (1999). https://doi.org/10.1016/S0263-8223(00)00031-3.
5. Prasanth S., Abhilash V., Kumaraswamy K.A.K., Thazhavilai P.D.R, Uma Thanu S.P. and Bellambettu C.P., Sliding wear behavior of Stir Cast AZ91/SiCp composites. *Journal of Solid Mechanics and Materials Engineering* 7(2), 169–175 (2013). https://doi.org/10.1299/jmmp.7.169.
6. Saravanan, R.A. Surappa M.K., Fabrication and characterisation of pure magnesium-30 vol.% SiCp particle composite. *Materials Science and Engineering* 276 (1–2), 108–116 (2000). https://doi.org/10.1016/S0921-5093(99)00498-0.
7. Deng, K. K., Wu, K., Wu, Y. W., Nie, K. B., and Zheng, M. Y. (2010). Effect of submicron size SiC particulates on microstructure and mechanical properties of AZ91 magnesium matrix composites. *Journal of Alloys and Compounds* 504(2), 542–547.
8. Poddar, P., Srivastava, V. C., De, P. K., Sahoo, K. L. Processing and mechanical properties of SiC reinforced cast magnesium matrix composites by stir casting process. *Materials Science and Engineering: A* 460, 357–364 (2007). https://doi.org/10.1016/j.msea.2007.01.052.
9. Kumar P.M., Balamurugan K., Uthayakumar M., Kumaran S.T., Slota A., Zajac J. (2019). Potential studies of waterjet cavitation peening on surface treatment, fatigue and residual stress. In: Gapiński B., Szostak M., Ivanov V. (eds) *Advances in Manufacturing II. Manufacturing 2019. Lecture Notes in Mechanical Engineering.* Springer, Cham, 978-3-030-16943-5. https://doi.org/10.1007/978-3-030-16943-5_31.
10. Ozdemir, I. and Toparli, M. An investigation of Al-SiCp composites under thermal cycling. *J Journal of Composite Materials* 37, 1839–1850 (2003).
11. Ghosh, D., Doloi, B., and Das, P.K., Parametric analysis and optimisation on abrasive water jet cutting of silicon nitride ceramics. *Journal of Precision Technology* 5(3/4), 294–311 (2015).
12. Chinnamahammad, B.A., Balamurugan, K., Fabrication and property evaluation of Al 6061 +x% (RHA +TiC) hybrid metal matrix composite. *SN Applied Science* 1, 977 (2019). https://doi.org/10.1007/s42452-019-1016-0.
13. Garikapati, P., Balamurugan, K., Latchoumi, T.P. et al. A cluster-profile comparative study on machining AlSi7/63% of SiC hybrid composite using agglomerative hierarchical clustering and K-means. *Silicon* (2020). https://doi.org/10.1007/s12633-020-00447-9.
14. Chua, B.W., Lu, L., and Lai, M.O., Infuence of SiC particles on mechanical properties of Mg based composite. *Composite Structures* 47, 595–601 (1999).

15. Sujayakumar, P., Viswanath, A., Kumar, K. K. A., Rajan, T. P. D., Pillai, U. T. S., and Pai, B. C., Sliding Wear behavior of stir cast AZ91/SiCp composites. *Journal of Solid Mechanics and Materials Engineering* 7(2), 169–175 (2013).

16. Poovazhagan, L., Rajkumar, K., Saravanamuthukumar, P., Javed Syed Ibrahim, S., and Santhosh, S., Effect of magnesium addition on processing the Al-0.8 Mg-0.7 Si/SiCp metal matrix composites. *Applied Mechanics and Materials* 787, 553–557 (2015). Trans Tech Publications Ltd.

17. Nie, K. B., Wang, X. J., Xu, F. J., Wu, K., and Zheng, M. Y,. Microstructure and tensile properties of SiC nanoparticles reinforced magnesium matrix composite prepared by multidirectional forging under decreasing temperature conditions. *Materials Science and Engineering: A* 639, 465–473 (2015).

18. Matin, A., Saniee, F. F., and Abedi, H. R., Microstructure and mechanical properties of Mg/SiC and AZ80/SiC nano-composites fabricated through stir casting method. *Materials Science and Engineering: A* 625, 81–88 (2015).

19. Bronfin, B., Katsir, M., and Aghion, E., Preparation and solidification features of AS21 magnesium alloy. *Materials Science and Engineering: A* 302(1), 46–50 (2001).

20. Szaraz, Z., Trojanova, Z., Cabbibo, M., and Evangelista, E., Strengthening in a WE54 magnesium alloy containing SiC particles. *Materials Science and Engineering: A* 462(1–2), 225–229 (2007).

21. Viswanath, A., Dieringa, H., Kumar, K. A., Pillai, U. T. S., and Pai, B. C., Investigation on mechanical properties and creep behavior of stir cast AZ91-SiCp composites. *Journal of Magnesium and Alloys* 3(1), 16–22 (2015).

22. Virtanen, P., Gommers, R., Oliphant, T. E., Haberland, M., Reddy, T., Cournapeau, D.,... and van der Walt, S. J., SciPy 1.0: fundamental algorithms for scientific computing in Python. *Nature Methods* 1–12 (2020).

23. Wang, S., Gittens, A., and Mahoney, M. W., Scalable kernel K-means clustering with Nyström approximation: relative-error bounds. *The Journal of Machine Learning Research* 20(1), 431–479 (2019).

24. Wu, M., Li, X., Liu, C., Liu, M., Zhao, N., Wang, J. and Zhu, L. Robust global motion estimation for video security based on improved k-means clustering. *Journal of Ambient Intelligence and Humanized Computing* 10(2), 439–448 (2019).

25. Jothi, R., Mohanty, S. K., and Ojha, A., DK-means: A deterministic k-means clustering algorithm for gene expression analysis. *Pattern Analysis and Applications* 22(2), 649–667 (2019).

26. Zhu, Q., Pei, J., Liu, X., and Zhou, Z., Analyzing commercial aircraft fuel consumption during descent: A case study using an improved K-means clustering algorithm. *Journal of Cleaner Production* 223, 869–882 (2019).

27. Yuan, C., Yang, H., Research on K-value selection method of K-means clustering algorithm. *Multidisciplinary Scientific Journal* 2(2), 226–235 (2019).

28. Virtanen, P., Gommers, R., Oliphant, T. E., Haberland, M., Reddy, T., Cournapeau, D., and van der Walt, S. J., SciPy 1.0: fundamental algorithms for scientific computing in Python. *Nature Methods* 17(3), 261–272 (2020).

29. Bauckhage, C., Numpy/scipy recipes for data science: k-medoids clustering. Researchgate. Net (2015).

30. Pedregosa, F., Varoquaux, G., Gramfort, A., Michel, V., Thirion, B., Grisel, O.,... and Vanderplas, J., Scikit-learn: Machine learning in Python. *The Journal of machine Learning Research* 12, 2825–2830 (2011).

31. Müllner, D., fastcluster: Fast hierarchical, agglomerative clustering routines for R and Python. *Journal of Statistical Software* 53(9), 1–18 (2013).

32. Douzas, G., Bacao, F., and Last, F., Improving imbalanced learning through a heuristic oversampling method based on k-means and SMOTE. *Information Sciences* 465, 1–20 (2018).

33. Capo, M., Perez, A., and Lozano, J. A., An efficient approximation to the K-means clustering for massive data. *Knowledge-Based Systems* 117, 56–69 (2017).
34. Al-Dabooni, S., Wunsch, D., Model order reduction based on agglomerative hierarchical clustering. *IEEE Transactions on Neural Networks and Learning Systems* 30:1881–1895 (2018).
35. Fernandez, A., Gomez, S., Versatile linkage: a family of space-conserving strategies for agglomerative hierarchical clustering. *Journal of Classification* 1–14 (2019).
36. Li, K., Yang, R. J., Robinson, D., Ma, J., and Ma, Z., An agglomerative hierarchical clustering-based strategy using Shared Nearest Neighbours and multiple dissimilarity measures to identify typical daily electricity usage profiles of university library buildings. *Energy* 174, 735–748 (2019).
37. Ranjan, M., Ivan, and Ramler, A k-mean-directions algorithm for fast clustering of data on the sphere. *Journal of Computational and Graphical Statistics* 19, 1–21 (2010).
38. Nguyen, Thuy-Diem, Bertil Schmidt, and Chee-Keong Kwoh. SparseHC: A memory-efficient online hierarchical clustering algorithm. *Procedia Computer Science* (2014). 10.1016/j.procs.2014.05.001.
39. Dabbabi, K., Hajji, S. and Cherif, A., Real-time implementation of speaker diarization system on raspberry PI3 using TLBO clustering algorithm. *Circuits System Signal Process* 39, 4094–4109 (2020). https://doi.org/10.1007/s00034-020-01357-2.
40. Gharibi, G., Alanazi, R., and Lee, Y., Automatic hierarchical clustering of static call graphs for program comprehension. In 2018 *IEEE International conference on big data (Big Data)* (pp. 4016–4025). IEEE (2018).
41. Park, H., Kwon, K., Khiati, A. I. Z., Lee, J., and Chung, I. J., Agglomerative hierarchical clustering for information retrieval using latent semantic index. In 2015 *IEEE International Conference on Smart City/SocialCom/SustainCom (SmartCity)* (pp. 426–431). IEEE (2015).
42. Cochez, M., Neri, F., Scalable hierarchical clustering: Twister tries with a posteriori trie elimination. In 2015 *IEEE Symposium Series on Computational Intelligence* (pp. 756–763). IEEE (2015).
43. Feltwell, T., Cielniak, G., Dickinson, P., Kirman, B. J., and Lawson, S., Dendrogram visualization as a game design tool. In *Proceedings of the* 2015 *Annual Symposium on Computer-Human Interaction in Play* (pp. 505–510) (2015).

9 Experimental Investigation on Interfacial Microstructure and Mechanical Properties of Al7075- White Bubble Alumina Syntactic Form

G. Rajyalakshmi, Jayakrishna Kandasamy, and S. K. Ariful Rahaman
Vellore Institute of Technology

G. Ranjith Kumar
Sri Venkateswara College of Engineering

A. Deepa
Vellore Institute of Technology

CONTENTS

DOI: 10.1201/9781003345466-9

167

9.1 INTRODUCTION

Reducing the weight of vehicles is an effective measure for energy saving in the automobile industry. Lightweight metals such as aluminium have widely replaced steel parts in automobile structures. The main purpose of this project is to prepare a metal matrix composite of high hardness and compressive strength for using it in aerospace and ship building industries.

Aluminium matrix with foam composites are getting demanded for many applications in recent periods. Many of the researchers used Al_2O_3 powder [1–3], hollow spheres of SiC [4, 5], fly ash [6, 7], glass cenospheres and different types of ceramic hollow particles [8, 9] as the foam in other aluminium alloys. Specially, commercial Al alloys have got a wide range of acceptance from industries due to their mechanical and dense properties. For corrosion resistance, weldability and machining performance, Bazarnik, et al. and Wen et al. [10, 11] suggested developing Al with Mg alloys. Pure Al is limited for brazing, bonding, liquid phase sintering or infiltration because of exhibition of obtuse contact angles on high melting point ceramics [12–14]. Interfacial reactions in ceramics were reported by Zhong et al. [15] and Li et al. [16].

Mechanical properties were sometimes controlled by the reaction occurred between the matrix and reinforcement. Especially for cellular materials, compression behaviour is one of the important properties for evaluating basic mechanical properties. Due to this many researchers [2] [8, 17, 18] studied the compressive properties of Aluminium Matrix Syntactic Foams (AMSFs) to understand the impact of energy absorption capacity on compressive behaviour. Luong [19] and Alvandi et al. [20] studied the relation between strain rate and compressive nature of AMSFs. It was mentioned that resistance of entrapped air inside the spheres may be attributed to the strain rate sensitivity of the base metal. Shear fracture [21] was observed in aluminium composites due to mass of brittle cenospheres, but the behaviour of compressive deformation is different with respect to the mechanical properties. Some of the authors [22] focused on investigating the influence of size and volume fractions of powder particles on the compression properties. After all the studies [23], it was proved that, the effect of interfacial bonding is inconsiderable on the compression behaviour of foam-based composites. When it comes to comparison, common aluminium foam consisting of gas porosity have shown better mechanical properties and superior thermal resistance and for better dimensional stability with low thermal expansion coefficient [24–26]. Owing with these properties, synthetic foam-based aluminium composites [27–28] are demanded as attractive material for many applications in aeronautical and automotive industries, which needs material with core structures for high damping applications.

Unlu [29] studied the properties of Al-based Al_2O_3 and SiC particle reinforced composite materials and found that mechanical properties like hardness of the composites significantly improved by the use of reinforcements. Veeresh Kumar [30] analysed the mechanical properties of Al6061-Al_2O_3 and Al7075-SiC composites and found that brinell hardness of the composites were increased with increase in filler content in the composites; the dispersion of Al_2O_3 in Al6061 and SiC in Al7075 alloy confirmed enhancement of the mechanical properties. Alumina (Al_2O_3) is an important structural ceramic material that is widely used in industry for various applications due to its excellent mechanical properties such as eight high mechanical performance, good chemical stability and high temperature characteristics. Bhaskar Chandra Kandpal [31] identified that Al_2O_3 particles were properly bonded to the aluminium matrix. The reinforcement of Al203 particles improved the microhardness and ultimate tensile strength of Aluminium Metal Matric Composites (AMMC). Kamat et al. (1989) [32] evaluated the mechanical properties of Al2024/Al_2O_3 composite and observed that yield and ultimate tensile strength of the composite increased with increase in volume fraction of Al_2O_3 particles. The author observed that yield strength of the composite increased while ultimate tensile strength and ductility decreased with increase in volume percentage of ceramic material. It was developed in situ Al-TiB2 composite by stir casting. They observed that the tensile and the yield strength of the composite was twice that of unreinforced matrix but the ductility showed a lower value.

The current research was motivated by the lack of information on Al-Al_2O_3 synthetic composites which are in high need for aerospace and ship building industries. Hence it is of great pleasure to carry out the research on Al synthetic foam-based composites and to improve the understanding of synthetic composites. Hence, the main objective of this work is to investigate the influence of addition of Al_2O_3 synthetic foam on microstructural changes produced by stir casting, and also to study the mechanical properties of Al-Al_2O_3 synthetic foam-based composites and to develop standard data, which could be referred to evaluate materials in future.

9.2 EXPERIMENTATION

9.2.1 MATERIAL

Al7075 was used as metal matrix composite and white bubble alumina was added as reinforcement to prepare composites in this study. Al_2O_3 is an electrical insulator but has a relatively high thermal conductivity (30 W/mK1) for a ceramic material. Aluminium oxide is insoluble in water. Alumina bubble is an electric arc furnace product made from high purity Bayer process alumina and the product chemistry is similar to that of white aluminium oxide. Bauxite is converted to aluminium oxide (Al_2O_3) by the Bayer process.

Relevant chemical equations are:

$$Al_2O_3 + 2NaOH \rightarrow 2NaAlO_2 + H_2O$$

$$2H_2O + NaAlO_2 \rightarrow Al(OH)_3 + NaOH$$

The intermediate, sodium aluminate, with the simplified formula $NaAlO_2$, is soluble in strongly alkaline water, and the other components of the ore are not. Depending on the quality of the bauxite ore, twice as much waste ("Bauxite tailings") as alumina is generated. Alumina bubbles are formed from molten alumina, resulting in hollow spheres of low bulk density. Bubble alumina is used in the production of lightweight insulating refractories where low thermal conductivity and

TABLE 9.1
Details of Sample Preparation

S. No.	% Al7075	% White bubble Alumina
1	95	5
2	90	10
3	85	15

high temperature properties are the prime requirements. It is also used effectively for loose-fill refractories.

The chemical composition of Al7075 used as matrix material roughly includes 5.6%–6.1% zinc, 2.1%–2.5% magnesium, 1.2%–1.6% copper, and less than a half percent of silicon, iron, manganese, titanium, chromium, and other metals. To increase the wettability of alumina particles in the molten Al, 1 wt% of magnesium (Mg) was added to molten aluminium during casting [30].

9.2.2 Fabrication of Al-Al$_2$O$_3$ Synthetic Foam

Alumina-reinforced Aluminium Matrix Composites (AMCs) were prepared in different compositions by stir casting process. Al7075 was melted in furnace and when the temperature of the liquid Al reached at 750°C, Mg was added in the melt. Alumina particles were added in molten metal through funnel at 730°C. An electrical resistance furnace assembled with graphite impeller used as stirrer was used for stirring purpose. After alumina addition, the liquid metal-reinforcements mixture was stirred for ten minutes. Finally composites were poured in metal moulds at 670°C. The melt was allowed to solidify in the mould. Three different samples (Table 9.1), each weighing 1 kg of Al7075-white bubble alumina with wt% of 95-5, 90-10, and 85-15, were prepared using the stir casting process.

9.3   RESULTS AND DISCUSSIONS

9.3.1 Microstructural Analysis

Samples were cut and polished using different grades of emery sheets and then grinded using grinding machine. Then etching is done using Keller's reagent. Under optical microscope, the microstructures of each sample are observed as shown in Figure 9.1 and SEM images are presented in Figure 9.2.

9.3.2 Hardness Test

The cube specimens of each sample are taken and performed Rockwell hardness test to determine its hardness. Rockwell hardness test is carried out on each specimen for three times and the average of those values is considered as the Hardness Rockwell B (HRBI) number of the matrix composite. Table 9.2 shows the values of HRB numbers of the specimens.

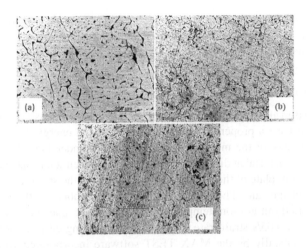

FIGURE 9.1 Microstructure images of samples. (a) Sample with 95-5. (b) Sample with 90-10. (c) Sample with 85-15.

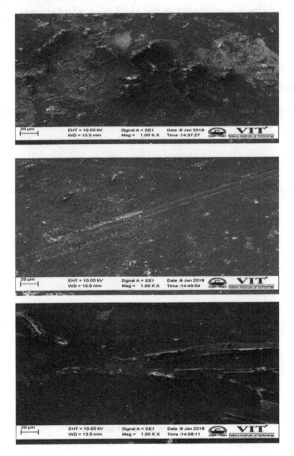

FIGURE 9.2 SEM images of samples. (a) Sample with 95-5. (b) Sample with 90-10. (c) Sample with 85-15.

From Table 9.2, it can be observed that the hardness of matrix composite is highest when alumina composition is 5% and then decreases gradually with increase in the wt% of alumina (Figure 9.3).

9.3.3 COMPRESSION TEST

Aluminium-based alloys usually possess application mostly in crash worthiness areas where material have a property of absorbing mechanical energy. In order to know the crashworthiness of the material, it is necessary to understand the compressive behaviour of the material at different strain rates. The specimen is placed in between the top and bottom plate of the setup such that the axis of the specimen is concentric with the axis of the ram. Then the hydraulic load is applied on the test specimen and test is carried out at room temperature. For each test, one specimen was taken and deformed to 0.05/s strain rate. The loads used during each deformation were recorded automatically by the MAX TEST software incorporated with the UTM machine. Compressive test was carried out at a strain rate of 0.05/s. Compression test is performed on two specimens of each sample and the average of the two values of each matrix composite is considered as the yield load, yield strength of it.

TABLE 9.2
Rockwell Hardness Values

Sample Number	Composition of Al-Al$_2$O$_3$	HRB-Scale Hardness Number
1	100-0	37.7
2	95-5	79.87
3	90-10	73.34
4	85-15	67.27

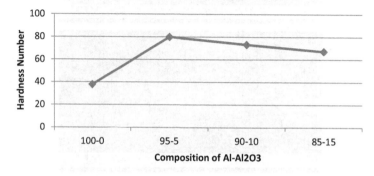

FIGURE 9.3 Hardness values of Al-Al$_2$O$_3$ composites with various compositions.

Typical load-displacement curve of aluminium synthetic foam is shown in Figure 9.4. Two typical phases were observed during compression. It starts with elastic phase and is defined as Young's modulus. This section is followed by plastic deformation phase where the load transfer between matrix and foam reaches its maximum. This is where the compressive strength is measured. Next, there is a small reduction in stress due to the reduced load-bearing capacity caused by the crushing of the ceramic micro-spheres, which results in the movement of the specimen. The second phase occurs between strains characterised by a relatively constant loads, where the micro-porosity in ceramic micro-spheres densifies plastically and where the fracture band expands. The energy absorption capacity can be found in this region, where the stress remains constant with the increasing strain (Figures 9.5 and 9.6).

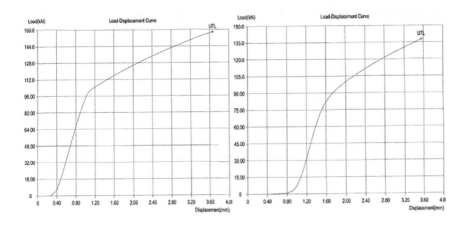

FIGURE 9.4   Load-displacement curve of Al-Al$_2$O$_3$ composition 95-5 specimen. (a) 1 and (b) 2.

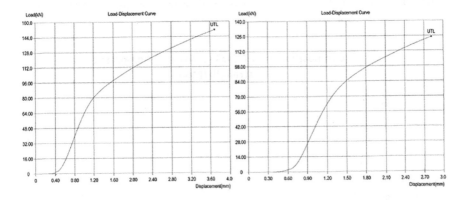

FIGURE 9.5   Load-displacement curve of Al-Al$_2$O$_3$ composition 90-10 specimen. (a) 1 and (b) 2.

Similar to the hardness of composites, compression load and compression strength of them are also highest when alumina composition is 5% and those values decrease gradually with increase in the wt% of alumina (Figure 9.7).

9.3.4 TENSILE TEST

Tensile test is performed on specimens of each sample and values of each matrix composite are considered as the yield load and yield strength of it (Figures 9.8 and 9.9).

9.3.5 IMPACT TEST

The impact test was carried out, and the following results are obtained in Table 9.3.

Thus, from the table, it is seen that the impact strength increases with the % composition of alumina foam in aluminium alloy (Figure 9.10).

From the graph, we can conclude that the tensile strength increases with the % composition of the alumina foam.

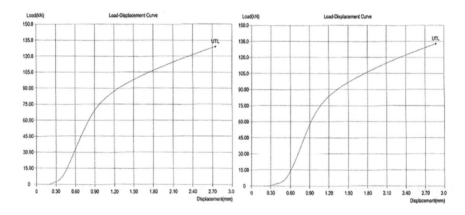

FIGURE 9.6   Load-displacement curve of Al-Al$_2$O$_3$ composition 85-15 specimen. (a) 1 and (b) 2.

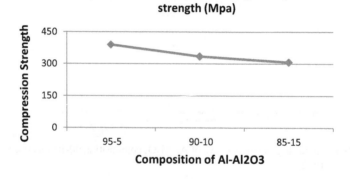

FIGURE 9.7   Comparison of compression strength of Al-Al$_2$O$_3$ composites.

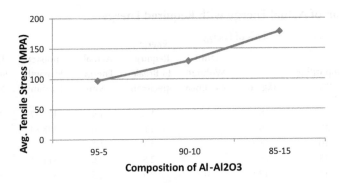

FIGURE 9.8 Comparison of tensile strength of Al-Al$_2$O$_3$ composites.

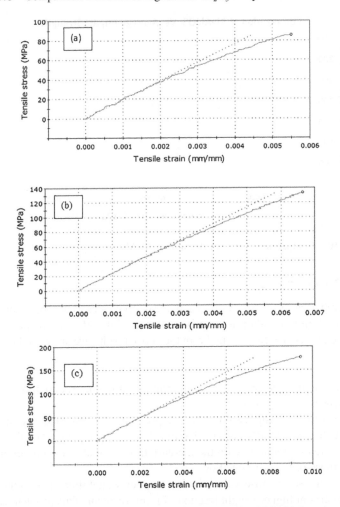

FIGURE 9.9 Stress-strain images of samples. (a) Sample with 95-5. (b) Sample with 90-10. (c) Sample with 85-15.

TABLE 9.3

Comparison of Actual Energy with Required Energy

S. No.	Composition (%)	(kg m)	Loss of Energy (Nm) Without Specimen	Energy Require to Break Specimen	Actual Energy (Nm)	Impact Strength (N/mm²)	Avg. Impact Strength (N/mm²)
1	5	8		1.6	1.4	177.77	177.77
2	10	9		1.8	1.6	188.84	
3	10	8.5		1.7	1.5	200	194.44
4	15	14		2.8	2.6	311.15	
5	15	13.5	0.2	2.7	2.5	300	305.55

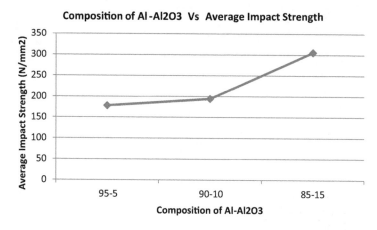

FIGURE 9.10   Comparison of impact strength of Al-Al$_2$O$_3$ composites.

9.4   CONCLUSIONS

This research presents a study of mechanical properties (hardness, compressive strength) of aluminium-alumina matrix composites of different compositions prepared by the process of stir casting. From the above results, we arrive at the following conclusions:

- It is economical to prepare an aluminium-alumina matrix composite using the process of stir casting.
- The hardness, compression load, compression strength suddenly increase when alumina of wt% 5 is added to the aluminium 7075 and then those values decrease gradually with the increase in wt% of alumina but are higher than those of the original alloy.
- The decrease in those values is due to poor wettability of the phases in the matrix at higher weight fraction of reinforcement. This problem can be

eradicated by adding sufficient amount of magnesium and by preheating the composites and the die.

- The mechanical properties of aluminium-alumina wt% 95-5 are closer to aluminium 7075-T6, but preparing aluminium 7075-T6 is not economical and takes more than 30 hours of time for preparing it.
- By using the process of stir casting, aluminium-alumina matrix composite can be prepared in less than three hours and also in an economical way.
- The tensile strength increases with the composition of the foam.
- The impact strength of the aluminium increases with the alumina foam composition.

REFERENCES

1. Kiser, M., He, M. Y., & Zok, F. W. (1999). The mechanical response of ceramic micro-balloon reinforced aluminum matrix composites under compressive loading. *Acta Materialia, 47*(9), 2685–2694.
2. Wang, H., Zhou, X. Y., Long, B., & Liu, H. Z. (2013). Compression behavior of Al2O3k/ Al composite materials fabricated by counter-gravity infiltration casting. *Materials Science and Engineering: A, 582,* 316–320.
3. Luong, D. D., Strbik III, O. M., Hammond, V. H., Gupta, N., & Cho, K. (2013). Development of high performance lightweight aluminum alloy/SiC hollow sphere syntactic foams and compressive characterization at quasi-static and high strain rates. *Journal of Alloys and Compounds, 550,* 412–422.
4. Cox, J., Luong, D. D., Shunmugasamy, V. C., Gupta, N., Strbik, O. M., & Cho, K. (2014). Dynamic and thermal properties of aluminum alloy A356/Silicon carbide hollow particle syntactic foams. *Metals, 4*(4), 530–548.
5. Rohatgi, P. K., Kim, J. K., Gupta, N., Alaraj, S., & Daoud, A. (2006). Compressive characteristics of A356/fly ash cenosphere composites synthesized by pressure infiltration technique. *Composites Part A: Applied Science and Manufacturing, 37*(3), 430–437.
6. Rohatgi, P. K., Gajdardziska-Josifovska, M., Robertson, D. P., Kim, J. K., & Guo, R. Q. (2002). Age-hardening characteristics of aluminum alloy-hollow fly ash composites. *metallurgical and Materials Transactions A, 33*(5), 1541–1547.
7. Balch, D. K., O'Dwyer, J. G., Davis, G. R., Cady, C. M., Gray III, G. T., & Dunand, D. C. (2005). Plasticity and damage in aluminum syntactic foams deformed under dynamic and quasi-static conditions. *Materials Science and Engineering: A, 391*(1–2), 408–417.
8. Tao, X. F., Zhang, L. P., & Zhao, Y. Y. (2009). Al matrix syntactic foam fabricated with bimodal ceramic microspheres. *Materials & Design, 30*(7), 2732–2736.
9. Xia, X., Chen, X., Zhang, Z., Chen, X., Zhao, W., Liao, B., & Hur, B. (2014). Compressive properties of closed-cell aluminum foams with different contents of ceramic microspheres. *Materials & Design (1980–2015), 56,* 353–358.
10. Bazarnik, P., Lewandowska, M., Andrzejczuk, M., & Kurzydlowski, K. J. (2012). The strength and thermal stability of Al–5Mg alloys nano-engineered using methods of metal forming. *Materials Science and Engineering: A, 556,* 134–139.
11. Wen, W., & Morris, J. G. (2003). An investigation of serrated yielding in 5000 series aluminum alloys. *Materials Science and Engineering: A, 354*(1–2), 279–285.
12. Ksiazek, M., Sobczak, N., Mikulowski, B., Radziwill, W., & Surowiak, I. (2002). Wetting and bonding strength in Al/Al2O3 system. *Materials Science and Engineering: A, 324*(1–2), 162–167.
13. Shen, P., Fujii, H., Matsumoto, T., & Nogi, K. (2004). Reactive wetting of SiO 2 substrates by molten Al. *Metallurgical and Materials Transactions A, 35*(2), 583–588.

14. Laurent, V., Chatain, D., & Eustathopoulos, N. (1991). Wettability of SiO2 and oxidized SiC by aluminium. *Materials Science and Engineering: A*, *135*, 89–94.

15. Zhong, W. M., L'esperance, G., & Suery, M. (2002). Effect of current Mg concentration on interfacial reactions during remelting of Al–Mg (5083)/Al2O3p composites. *Materials Characterization*, *49*(2), 113–119.

16. Guobin, L., Jibing, S., Quanmei, G., & Yuhui, W. (2005). Interfacial reactions in glass/Al–Mg composite fabricated by powder metallurgy process. *Journal of Materials Processing Technology*, *161*(3), 445–448.

17. Mondal, D. P., Das, S., Ramakrishnan, N., & Bhasker, K. U. (2009). Cenosphere filled aluminum syntactic foam made through stir-casting technique. *Composites Part A: Applied Science and Manufacturing*, *40*(3), 279–288.

18. Wu, G. H., Dou, Z. Y., Sun, D. L., Jiang, L. T., Ding, B. S., & He, B. F. (2007). Compression behaviors of cenosphere–pure aluminum syntactic foams. *Scripta Materialia*, *56*(3), 221–224.

19. Luong, D. D., Gupta, N., Daoud, A., & Rohatgi, P. K. (2011). High strain rate compressive characterization of aluminum alloy/fly ash cenosphere composites. *JOM*, *63*(2), 53–56.

20. Alvandi-Tabrizi, Y., Whisler, D. A., Kim, H., & Rabiei, A. (2015). High strain rate behavior of composite metal foams. *Materials Science and Engineering: A*, *631*, 248–257.

21. Myers, K., Katona, B., Cortes, P., & Orbulov, I. N. (2015). Quasi-static and high strain rate response of aluminum matrix syntactic foams under compression. *Composites Part A: Applied Science and Manufacturing*, *79*, 82–91.

22. Santa Maria, J. A., Schultz, B. F., Ferguson, J. B., Gupta, N., & Rohatgi, P. K. (2014). Effect of hollow sphere size and size distribution on the quasi-static and high strain rate compressive properties of Al-A380–Al2O3 syntactic foams. *Journal of Materials Science*, *49*(3), 1267–1278.

23. Orbulov, I. N., & Májlinger, K. (2013). Description of the compressive response of metal matrix syntactic foams. *Materials & Design*, *49*, 1–9.

24. Orbulov, I. N. (2012). Compressive properties of aluminium matrix syntactic foams. *Materials Science and Engineering: A*, *555*, 52–56.

25. Rohatgi, P. K., Gupta, N., & Alaraj, S. (2006). Thermal expansion of aluminum–fly ash cenosphere composites synthesized by pressure infiltration technique. *Journal of Composite Materials*, *40*(13), 1163–1174.

26. Cox, J., Luong, D. D., Shunmugasamy, V. C., Gupta, N., Strbik, O. M., & Cho, K. (2014). Dynamic and thermal properties of aluminum alloy A356/Silicon carbide hollow particle syntactic foams. *Metals*, *4*(4), 530–548.

27. Palmer, R. A., Gao, K., Doan, T. M., Green, L., & Cavallaro, G. (2007). Pressure infiltrated syntactic foams—Process development and mechanical properties. *Materials Science and Engineering: A*, *464*(1–2), 85–92.

28. Goel, M. D., Matsagar, V. A., & Gupta, A. K. (2015). Blast resistance of stiffened sandwich panels with aluminum cenosphere syntactic foam. *International Journal of Impact Engineering*, *77*, 134–146.

29. Bekir Sadik Unlu. (2008). Investigation of tribological and mechanical properties Al2O3- SiC reinforced Al composites manufactured by casting or P/M method. *Proceedings of Materials & Design*, *29*(10).

30. Bhaskar Chandra Kandpal, Jatinder kumar, Hari Singh. (2016). Fabrication and characterisation of Al2O3/aluminium alloy 6061 composites fabricated by Stir casting. *ICMPC*.

31. Veeresh Kumar, G. B., Rao, C. S. P., Selvaraj, N., & Bhagyashekar, M. S. (2010). Studies on Al6061-SiC and Al7075-Al2O3 metal matrix composites. *Proceedings of JMMCE*, *9*.

32. Annigeri, U. K. & Veeresh Kumar G. B. (2016). Method of stir casting of aluminum metal matrix composites: A review. *Journal of ICMPC*.

10 Investigation of the Humping Phenomenon in Wire Arc Additive Manufacturing

Nobel Karmakar and Ritam Sarma
Indian Institute of Technology Guwahati, India

Ranjeet Kumar Bhagchandani
Vellore Institute of Technology

Sajan Kapil
Indian Institute of Technology Guwahati, India

CONTENTS

10.1 INTRODUCTION

Additive manufacturing (AM) is one of the integral parts of Industry 4.0 because of its capability to manufacture any component with total automation. Developed in the 1980s, additive manufacturing refers the fabrication of three-dimensional object directly from the digital data, unlike conventional subtractive methods in which the component is obtained by removing the materials from raw stock. *AM*

DOI: 10.1201/9781003345466-10

has gained importance among researchers because of its ability to manufacture any customized and complex shape components with a high degree of design freedom, minimized wastage of materials, and functional integration. The manufacturing of complex metallic objects was considered the bottleneck of conventional manufacturing processes. Complex components need to be broken into small sub-assembly parts, which need to be manufactured separately by using conventional manufacturing processes. Later on, the parts are assembled in order to have the final component. Traditional manufacturing results in the wastage of metals as well as manufacturing lead time. Such difficulties can be overcome by shifting from traditional manufacturing to *metal additive manufacturing (MAM)* processes. *Wire arc additive manufacturing (WAAM)* has recently gained popularity among the industrial academia people due to its capability of high deposition rate in comparison with the other *MAM* processes. This technology enables people to build any large metallic components such as turbine blades and impellers in a brief time with less wastage of materials at a minimal cost.

WAAM uses an electric arc generated between the substrate and electrode to fuse feedstock material. In this process, the wire feedstock material is used. It can efficiently manufacture near-net-shaped components with short lead time and millimeter-scales resolution. Moreover, metal wire is cost-efficient compared to metal powder and is more readily available with appropriate properties for *AM*. Thus, this technology offers an alternative way to traditional machining and generate structures with complex features. In 1925, an American researcher named Baker proposed the notion of using the electric arc as a fusion source and filler wire as feedstock to create metallic ornaments [1]. But this technology has gained importance among researchers and has been developed for the past few years. The electric arc in this process melts the wire, after which the molten metal is transferred to the melt pool. The process planning for *WAAM* involves the orientation of the part relative to the deposition table, slicing the 3D models into 2D layers for generating the process building motions. Inappropriate process planning and deposition result in defects such as high residual stress, porosity, cracks, and poor surface finish affecting the product life of the part being fabricated through *WAAM* [2].

Multi-directional *WAAM* or multi-orientation *WAAM*, or positional deposition in *WAAM*, has gained much popularity these days as they possess the ability to fabricate complex as well as overhanging structures without any extra support. However, depositing overhanging structures also yields some defects in the form of the humping phenomenon, which can deteriorate the shape of the weld beads. Improper weld parameters, as well as Marangoni force, play a role in the formation of humped beads. Additionally, compared to the standard *gas metal arc welding (GMAW)*-based *WAAM* method, it is challenging to appropriately regulate the deposition process due to the downward flow of the molten metal caused by the gravitational pull. Free flight and short-circuiting modes can be used to categorize the main metal transfer types of *GMAW*. Spray, globular, and repelled transfer modes are further classified under free flight transfer. The droplet trajectory from the wire tip to the weld pool is crucial during free flight modes. These types of metal transfer are governed by various factors, including electromagnetic force, surface tension force, and gravity force.

The gravitational force may be strong enough in positional free flight transfer to overcome other factors that would ordinarily project the droplet across the arc and cause the material to be displaced from the desired position [3]. A droplet forms on the wire tip during the arcing phase of the short-circuit transfer mode, but material is only transferred when the wire tip makes contact with the base metal. The droplets are separated from the filler wire tip to the molten weld pool by surface tension and electromagnetic forces. Once the droplet is transferred, an arc gap is created, due to which the voltage increases abruptly, which is followed by the reignition of the arc; thus, the droplet forms, and the process repeats. There is less influence of the gravitational force on the transfer of droplets to the weld pool. With the correct combination of welding parameters, the accurate transfer of metal to the target position occurs. Because of this, the short-circuit mode is the preferred option for material deposition in all situations. The dynamic adjustment of welding current and other welding parameters can enhance the positioning performance of both free flight and short-circuiting transfer modes.

Gravitational pull causes the molten metal to flow downhill in a sagging bead, which is the fundamental problem with horizontal deposition. A bad bead profile might emerge from the molten pool sagging before it freezes during the solidification of the liquid metal. As a result, it is crucial to research how various forces in the multi-directional *GMAW*-based *WAAM* process affect the final bead quality. For the purpose of analyzing the geometry of the weld pool, a static force balance model for a molten pool in horizontal welding is used (Figures 10.1–10.3).

Gravitational force G, arc pressure f_{arc}, and surface tension force f_γ are additional factors that affect the development of the molten pool in addition to the normal force N. According to Randhawa's static force balancing hypothesis, the molten droplet will remain balanced on the base metal in the vertical position as long as the static detaching forces do not exceed the holding force. The surface tension force can be divided into horizontal and vertical components, according to the force model. While the gravity force tends to lead to the detachment of the pendent molten pool during

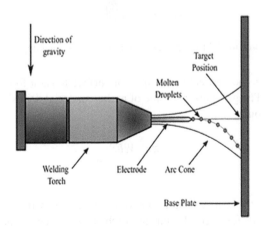

FIGURE 10.1   Trajectory of molten droplets in free flight transfer mode [3].

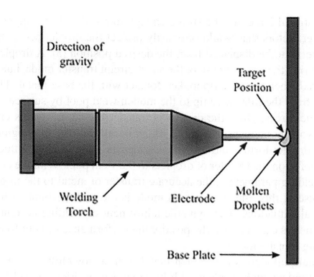

FIGURE 10.2 Trajectory of molten droplets in short-circuit transfer mode [3].

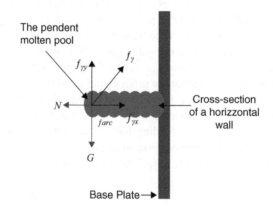

FIGURE 10.3 A force model for a pendant pool in the positional deposition [3].

horizontal welding, the vertical component of surface tension force f_γ serves as the holding force [3]. The gravitational force of molten material deposited in unit time is given by the relation:

$$G = \frac{\pi r^2 \, WFS}{TS} \, \rho g$$

where ρ and r are the density and radius of the electrode wire, respectively. WFS and TS are the wire feed speed, and the torch travel speed, respectively; g is the gravitational constant. Therefore, the gravitational force is inversely proportional to the torch travel speed. In general, the surface tension force, which operates on the curved liquid surface at the border of the liquid and solid material region, is the main force responsible for keeping the pendent molten pool during positional welding and allowing the molten bead to stay stable. An important element that influences this force is the molten metal's surface tension coefficient, which measures the surface tension force per unit length. The Marangoni force is one of the primary causes of metal convection in the weld pool. It is a sort of surface tension force that drives the flow of a molten metal stream from a low surface tension area to a high surface tension region. The surface tension coefficient γ is considered to vary linearly with the temperature T, according to the relation given by Xiong et al. [4]:

$$\gamma = \gamma_m^0 - A_s \, (T - T_m)$$

where γ_m^0 is the surface tension coefficient for pure metal at the melting point, which is a constant for filler metal. T_m is the melting point of the filler material. When the temperature is significantly over the melting point, As can be computed as the negative of for pure metal, which is also regarded as a constant. In order to benefit from surface tension rod retention, it is therefore preferable to have a lower temperature weld pool; in other words, the heat input should be as low as possible in order to increase the surface tension force. Additionally, the molten pool's freezing pace affects the way it takes on a shape. The pool will keep its shape if it freezes rapidly. A faster freezing rate is determined by a lower heat input [4]. Additionally, because surface tension is thought of as a temperature-dependent coefficient, the gradient of surface tension and weld pool temperature can be related. As a result, the metal stream has a tendency to go from a location of high temperature to one of relatively low temperature. The weld pool beneath the electrode in the welding process has a lower surface tension and a greater temperature than the weld pool's tail. As a result, the Marangoni force increases the momentum of the metal flow, which gives rise to the humping phenomenon resulting in humped beads.

In the multi-directional, multi-layer *GMAW*-based *WAAM* process, humping is a frequent flaw. It can be characterized as the weld bead's periodic undulation during high-speed welding. Inappropriate weld parameters, such as system energy input, welding speed, etc., can result in the humped weld bead. The molten pool initially forms and progresses to the molten pool growth phase. The surface of the molten pool beneath the wire electrode is lowered by the arc force, creating a strong metal flow. High momentum fluid flow causes the weld pool's tail to bend and slope as well as somewhat swell as it goes backward to it. The swelling keeps becoming bigger as the metal vapor keeps going toward the back of the molten pool. As the welding speed increases, the weld pool expands, as does the temperature differential between the front and tail of the molten pool [5]. This causes a significant Marangoni force and, consequently, an increase in the velocity of the metal flow (Figure 10.4).

Between the wire electrode and the humped bead, a lengthy liquid metal jet can be seen since the welding flame is still moving forward at a fast rate of speed. As the heat source advances ahead in the welding direction, the metal flow is stopped from returning to the swelling zone until a portion of the liquid jet hardens. A fresh

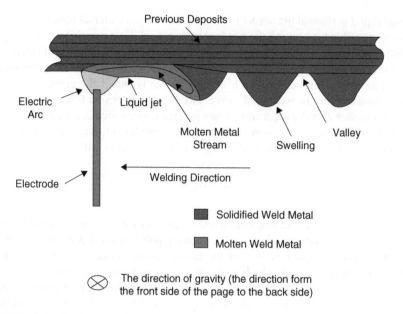

FIGURE 10.4 A schematic diagram of the humping phenomenon [12].

humping begins to develop when the earlier swelling stops expanding. Therefore, the powerful, high-velocity flow of molten metal is to blame for the humping creation during high-speed welding. However, as a result of the combined impact of droplet momentum, arc pressure, the electromagnetic force, and surface tension, a clockwise circulation and a backward pattern of fluid flow occur in the weld pool behind the arc in the typical speed *GMAW* process. The recirculation of molten metal toward the weld pool's front limits weld reinforcing. As a result, there is no buildup of molten metal in the weld pool's trailing edge [5]. Furthermore, the variation in weld width in the weld direction is negligible when using the regular speed *GMAW* technique. In normal speed *GMAW* deposition, the capillary pressure of the liquid channel does not vary significantly in the welding direction since it is proportional to the curvature in a cylindrical section. The liquid metal is forced from the weld pool center to the pool edge by the surface tension because the surface tension at the weld pool center is lower than that at the near edge. The little variation in weld width is a result of this fluid flow pattern that is directed outward. The flow pattern does not alter as the welding time increases in the high-speed *GMAW* process.

Molten metal flows to the back of the weld pool, but it does not backfill the front portion of the weld pool. When a result, as the welding time grows, the swelling's size keeps growing and the liquid channel lengthens. The amount of the molten metal in the liquid channel diminishes as the buildup of molten metal at the weld pool's trailing edge rises, which narrows the weld in the direction of the weld. The capillary pressure of the liquid channel at the front section of the weld pool is greater than that at the back end of the weld pool as a result of the transverse curvature variation along the liquid channel. The liquid channel's capillary pressure variation in the welding direction may encourage liquid channel shrinkage, which in turn slows the flow of molten metal between the leading and trailing edges of the weld pool. Additionally, the molten metal's temperature in the liquid channel decreases as a result of heat loss, which results in the melt's poor fluidity. These factors all play a part in the liquid channel's propensity to harden too soon. In order for the hump to form, the channel's neck must first solidify. This shows how the valley, which divides the weld pool into two sections, formed.

Gratzke et al. [6] applied Rayleigh's instability theory and discussed a theoretical approach to the humping phenomenon in welding processes. They considered the length-to-width ratio of the weld pool as the governing quantity, and it has to remain below a critical value for the humping to take place. Therefore, maximization of the length-to-width ratio is required to avoid humping. Humping was divided into two unique forms of formation by Soderstrom and Mendez [7], namely *gouging region morphology (GRM)* and *beaded cylinder morphology (BCM)*. Between the humped beads, there are open, empty dry patches that define the *GRM*. The front of the weld pool, meanwhile, shows a significant dip known as the "gouging zone." The back of the weld pool is where the majority of the trailing region, or molten metal, is located. On the other hand, the distinctive characteristics of *BCM* include bead-like protuberances that are joined by a small central channel and sit above the surface of the workpiece. Additionally, they talked about how different welding settings affected *GRM* and *BCM*. In their thorough investigation of high-speed weld bead flaws, Nguyen et al. [8] placed a lot of emphasis on the development of humping. They looked at numerous theories that were put up to explain the occurrence of hump formation, including the Rayleigh jet instability model, the Arc pressure model, the supercritical flow model, etc.

According to these models, humping was caused by fluid flow, arc pressure, metallostatic pressure, and the high velocity and high momentum of the backward-directed flow of molten metal in the weld pool. In order to prevent the *GMAW* humping problem, Choi et al. [9] looked into an unique hybrid laser beam welding plus *GMAW* technology. They reasoned that adding more heat sources to the welding process would make weld beads broader and lessen capillary instability. To prevent the development of weld bead hump flaws and enable faster travel rates, they applied heat input from a defocused laser beam in front of a *GMAW* melt pool. Kazanas et al. [10] fabricated inclined and horizontal wall features with an inclined torch position in *CMT* mode with varying travel speeds to understand the influence of travel speed on part quality for angled walls. They observed deteriorated wall quality at high travel speeds. In order to prevent the humping bead in the high-speed *GMAW* process,

Wang et al. [11] introduced an additional electromagnetic force that was produced by the interaction of an external magnetic field and the welding current to suppress the momentum of the rearward flowing metal. The progression of the molten metal toward the tail of the weld pool can be effectively slowed down by applying a supplemental electromagnetic field to the forepart of the weld pool. As a result, the momentum of the molten metal at the tail of the weld pool cannot overcome the surface tension and hence cannot accumulate to swelling, and the humping was suppressed. The stability of positional deposition was examined by Yuan et al. [12] in relation to process variables such as wire feed speed, torch travel speed, stand-off distance, and torch tilt. Because of the combined impacts of the prolonged metal freezing time, greater weld bead volume, and reduced surface tension force, they found that the bead shape degrades with increasing WFS. From the literature, it can be inferred that *WAAM* has the flexibility to deposit pieces with overhanging features in any direction without the need for extra support structures. This is due to the positioning capacity of various welding methods. The humping effect, which manifests as a series of periodic bead-like protuberances on the weld deposits, may cause the dimensional quality of the overhanging components to deteriorate. In conventional fusion welding methods, the production of humping is a common weld fault, especially at faster torch travel speeds.

Higher torch travel speeds result in a larger weld pool, a larger temperature differential between the front and back of the melt pool, a larger Marangoni force, and consequently more metal flow momentum. The fast-moving metal stream continues to flow backward, which causes the molten pool's rear end to hump. An elongated liquid metal jet emerges between the wire electrode and the hump while the torch keeps moving forward at a high rate of speed; as a result, the previous swelling stops expanding and a new hump starts to build. If humping can be avoided by devising a complex robot trajectory and employing the torch travel speed under the most constrained conditions, multi-directional *WAAM* will prove to be more advantageous.

10.2 MATERIAL AND METHODS

In this work, the effect of different parameters on formation of humping in multi-axis deposition has been studied experimentally. All the depositions were performed on a 6-axis robotic *WAAM* setup. Mild steel plates of dimensions $150 \times 100 \times 6.3$ mm were used as substrates, and ER70S-6 copper-coated wires of a diameter of 1.2 mm were used as the metallic feedstock material. Samples from the deposited horizontal walls were obtained by wire-cut EDM, after which those samples were thoroughly polished and grinded before subjecting to microhardness and microstructural analysis.

10.2.1 *WAAM* ROBOT SETUP

The experimental setup consists of a six-axis Fanuc robot ARC mate 100iD integrated with the Fronius TPS 400i *GMAW* unit, as shown in Figure 10.5. The robotic arm serves as a deposition head of *WAAM*. The deposition head is used as an end effector of the robotic arm, which holds the consumable electrode and maintains an arc between substrate and electrode. The whole setup consists of two parts motion module and a deposition module; the motion module consists of a robot controller which controls the motion of joints, as a result, controls deposition velocity. The deposition module consists of a deposition power source, shielding gas, and material spool, which controls the weld bead parameters by using the cold metal transfer type of material transfer process (Table 10.1).

10.2.2 MICRO VICKERS' HARDNESS TESTER SETUP

The Micro Vickers' Hardness Tester is a precise testing system suitable for hardness analysis of metallic specimens in metallography laboratories. The sample is placed

FIGURE 10.5 Experimental setup of robotic *WAAM*: 1. material deposition power source; 2. robotic arm; 3. deposition head; 4. deposition torch; 5. welding table; and 6. shielding gas cylinder.

TABLE 10.1
Specifications of the Cold Metal Transfer Unit

Machine Parameters	Remarks
Current	3–400 A
Open circuit voltage	73 V
Mains frequency	50–60 Hz
Shielding gas	80%–20% Ag-CO_2 mixture
Shielding gas flow rate	15 l/min
Operation mode	2 step modules
Duty cycle	100% at 320 A

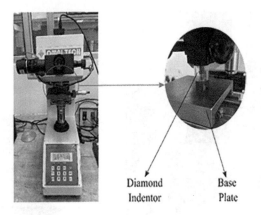

Diamond Indentor   Base Plate

FIGURE 10.6 Omni Tech Micro Vickers' Hardness Tester setup.

beneath the diamond indentor, and the dwell time and load are set. Then the indentor does its job, and the micro hardness value is calculated with the help of a suitable application on the computer (Figure 10.6 and Table 10.2).

10.2.3 Microscope Setup

The microscope that is utilized for capturing the microscopic images is Leica ICC500W. It provides excellent image sharpness, brightness, and color impression. In addition, the user also can work with the basic presets to adjust the camera's parameters as desired. The onboard buttons of the camera can be pressed to quickly switch the camera modes, perform white balance, or save an image to an SD card. The polished specimen is kept under the microscope, and then the lens is adjusted to acquire images of desired magnification. The base plate on which the specimen is kept is adjustable so that the microscopic images of various spots in the specimen can be captured without touching the specimen (Figure 10.7 and Table 10.3).

TABLE 10.2
Specifications of Omni Tech Micro Vickers' Hardness Tester

Parameters	Remarks
Load range	10–1000 g
Loading	Automatic
Dwell time	5–99 s
Microscopic lens magnification	10×, 40×
Measurement of indentation	Digital filar eyepiece
Maximum height of the specimen	65 mm
Base plate dimension	100 × 100 mm

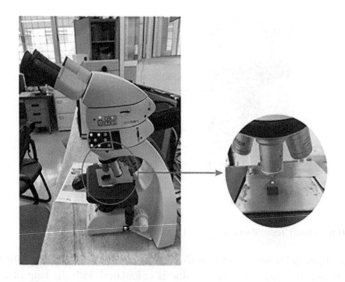

FIGURE 10.7   Leica ICC 500W Microscope setup.

TABLE 10.3

Specifications of Leica ICC 500W Microscope

Parameters	Remarks
Weight	700 g (camera only)
Height	50 mm
Exposure time	1500 msec
Live image	30 fps maximum
Microscopic lens magnification	5×, 10×, 50×
Movie clip	1920 × 1080 maximum
Color depth	24 bit

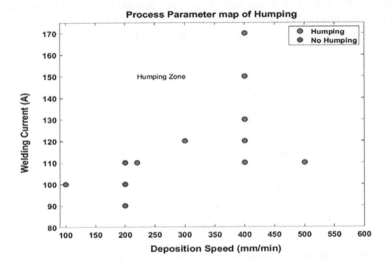

FIGURE 10.8 Identification of humping on process parameter map.

10.3 RESULTS AND DISCUSSIONS

A number of horizontal depositions were performed with varying process parameters, and the selection criteria were that those parameters had to show no humped beads up to five layers. Those combinations of parameters are represented in Figure 10.8. The red dots represent the experiments that displayed the occurrence of humped beads before or at the 5th layer, whereas the green dots were those that maintained their shape even after the 5th layer. Therefore, from the process parameter map, one fact that can be inferred is that a low welding current coupled with a lower value of torch travel speed, i.e., a lower energy input yields a weld bead shape with a desirable appearance. In this context, a welding current of 100 A at 100 mm/min should have produced the best weld bead, but while depositing, it was observed that due to the very high energy input of 762 J/mm, the fluidity of the molten metal was very high. Because of this reason, the molten weld pool tended to deform in the downward direction and had a deteriorated appearance.

10.4 GEOMETRICAL ANALYSIS OF MULTI-LAYER HORIZONTAL GMAW DEPOSITION UP TO THREE LAYERS

Single bead depositions were carried out horizontally with *GMAW* in CMT mode at varying welding parameters. Multi-layer deposition at varying welding currents and the same torch travel speed was carried out to comprehend the influence that current or energy input has on the quality of the bead appearance. The same torch travel speed of 400 mm/min was used, and the wire feed speed and stand-off distance were maintained constant at 2.9 m/min and 9.5 mm, respectively.

From Table 10.7, the main point that can be inferred is that energy input plays a significant role in determining the weld bead quality since the deposition at 110 A welding current has the best bead shape among the four depositions. Furthermore, a quantitative comparison of the difference between the average height of the humped beads and regular beads reflects that as the welding current reduces, the difference between the heights of the humped bead and the regular bead also reduces, i.e., lower the energy input, better is the weld bead shape (Figures 10.9–10.11 and Tables 10.4–10.6).

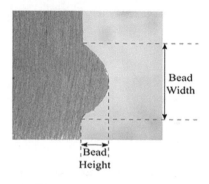

FIGURE 10.9   *GMAW* single bead profile at 110 A.

TABLE 10.4
Process Parameters for *GMAW* Single Bead

Parameters	Remarks
Welding current (A)	110
Welding voltage (V)	16.9
Torch travel speed (mm/min)	200
Wire feed speed (m/min)	2.9
Stand-off distance (mm)	9.5
Energy input (J/mm)	557.7

The bead height and width obtained were 1.58 and 4.41 mm, respectively.

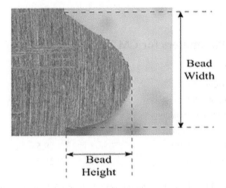

FIGURE 10.10 CMT single bead profile at 100 A.

TABLE 10.5
Process Parameters for CMT Single Bead 1

Parameters	Remarks
Welding current (A)	100
Welding voltage (V)	12.7
Torch travel speed (mm/min)	200
Wire feed speed (m/min)	2.9
Stand-off distance (mm)	9.5
Energy input (J/mm)	381

The bead height and width obtained were 3.34 and 6.99 mm, respectively.

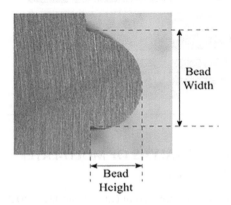

FIGURE 10.11 CMT single bead profile at 90 A.

TABLE 10.6

Process Parameters for CMT Single Bead 2

Parameters	Remarks
Welding current (A)	90
Welding voltage (V)	12.1
Torch travel speed (mm/min)	200
Wire feed speed (m/min)	2.9
Stand-off distance (mm)	9.5
Energy input (J/mm)	326.7

The bead height and width obtained were 2.86 and 5.28 mm, respectively.

TABLE 10.7

***GMAW* Multi-Layer Horizontal Depositions up to Three Layers with Varying Process Parameters**

Depositions	Welding Parameters	
	Welding Current (A)	Energy Input (J/mm)
	170	479.4
	150	398.25
	130	337.35
	110	278.85

10.5   GEOMETRICAL ANALYSIS OF MULTI-LAYER HORIZONTAL WALLS

A horizontal deposition was performed in *GMAW* mode at 110 A, and a total of seven layers were deposited. Up to five layers, there were very subtle signs of the appearance of humping, but from the 6th layer onwards, humping became more prominent, and after the 7th layer, the deposition could not be further proceeded. The other process parameters were the same as that for the *GMAW* single bead. The height and thickness of the wall obtained were 18.3 and 6.1 mm, respectively (Figure 10.12).

A horizontal deposition at a welding current of 100 A was performed in CMT mode, and a total of ten layers were deposited. Up to nine layers, there were very subtle signs of the appearance of humping, but from the 10th layer onwards, humping became more evident. The other process parameters were the same as that for the CMT single bead at 100 A. The height and thickness of the wall obtained were 24.6 and 6.7 mm, respectively (Figure 10.13).

A horizontal deposition at a welding current of 90 A was performed in CMT mode, and a total of ten layers were deposited. There were almost no signs of the appearance of humping even after depositing ten layers. The other process parameters were the same as that for the CMT single bead at 90 A. The height and thickness of the wall obtained were 24.4 and 6.3 mm, respectively (Figure 10.14).

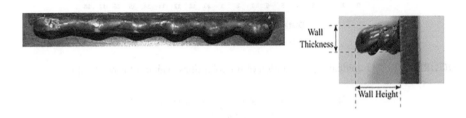

FIGURE 10.12   *GMAW* horizontal wall at 110 A up to seven layers.

FIGURE 10.13   CMT horizontal wall at 100 A up to ten layers.

FIGURE 10.14   CMT horizontal wall at 90 A up to ten layers.

10.6   MICROHARDNESS ANALYSIS

All the samples were put beneath the diamond indentor in the Micro Vickers' Hardness Tester, and a load of 500 g was set along with a dwell time of ten seconds. The average microhardness values in *GMAW* CMT samples at 100 and 90 A are 284.011, 177.62, 182.38 HV. The microhardness results are plotted in the graphs presented in Figures 10.15–10.18.

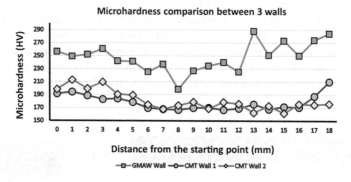

FIGURE 10.15   Graphical representation of microhardness values of three samples.

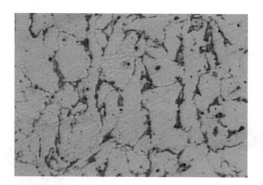

FIGURE 10.16   Microstructure of CMT Wall sample at 90 A.

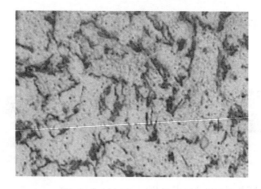

FIGURE 10.17   Microstructure of CMT Wall sample at 100 A.

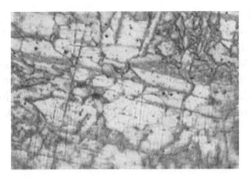

FIGURE 10.18   Microstructure of the *GMAW* Wall sample at 110 A.

10.7   MICROSTRUCTURE ANALYSIS

From the above pictorial representations, it can be inferred that the microstructure basically consists of ferrite and pearlite structures. Pearlite is a lamellar arrangement of cementite in ferrite. The pearlite phase can be identified as black spots segregated along the grain boundaries. The inner white areas represent the ferrite phase. In the case of the CMT 100A sample, this pearlite segregation is comparatively more than that of the CMT 90A sample. Segregation of pearlite results in the decrease of the hardness of the material as well as the strength. The microhardness results reflect this fact. Furthermore, the average grain size in the *GMAW* sample is lower in comparison to CMT samples. This reduced grain size is responsible for the increased hardness observed in the hardness test.

10.8   CONCLUSION

AM intends to change the vision of how the products will evolve and be realized. Multi-directional *WAAM* has the potential to fabricate complex shapes and improve product development efficiency as well as manufacturing execution. Robotic *WAAM* has the great advantage of a high deposition rate, but this boon comes with a bane in the form of challenges such as the humping phenomenon and an increased gravitational force in the positional deposition. Humping is the periodic occurrence of bead-like protuberances in high-speed welding. Marangoni force is the most significant factor responsible for the occurrence of the humping phenomenon. Experimental data shows that a lower value of current coupled with a lower torch travel speed suppresses the humping phenomenon. So, keeping this fact in mind, an attempt was made to perform horizontal deposition and three different walls, one in *GMAW* mode (up to seven layers) at a current of 110 A, one in CMT mode (up to ten layers) with a current of 100 A and the other in *CMT* mode (up to ten layers) but with a current of 90 A. The torch travel speed was kept constant at 200 mm/min. Samples from these walls were subjected to the Vickers Microhardness Test, where the microhardness of the *GMAW* sample was 28% and 26% more than that of the *CMT* samples at 100 and 90 A, respectively. This hardness behavior was explained by the microstructural analysis, where it was found that the average grain size in the *GMAW* sample was lower than that of the other samples. Moreover, CMT 100 A sample has more pearlite segregation than the CMT 90 A sample and thus has comparatively lower hardness. Furthermore, a BT40-based *GMAW-WAAM* torch holder has been developed to utilize the entire working volume of a retrofitted CNC machine. The tool can be directly mounted on the CNC tool holder. Hence, the common work coordinate system can be utilized for effective hybrid machining. Therefore, it can be concluded that a combination of low welding current and low torch travel speed, i.e., lower energy input, can suppress the humping phenomenon and improve the weld bead quality as well as the appearance.

REFERENCES

1. R. Baker, "Method of making Decorative Articles," *ACM SIGGRAPH Comput. Graph.*, vol. 28, no. 2, pp. 131–134, 1925, doi: 10.1145/178951.178972.
2. M. Jamshidinia, M. M. Atabaki, M. Zahiri, S. Kelly, A. Sadek, and R. Kovacevic, "Microstructural modification of Ti-6Al-4V by using an in-situ printed heat sink in Electron Beam Melting® (EBM)," *J. Mater. Process. Technol.*, vol. 226, pp. 264–271, 2015, doi: 10.1016/j.jmatprotec.2015.07.006.
3. L. Yuan et al., "Fabrication of metallic parts with overhanging structures using the robotic wire arc additive manufacturing," *J. Manuf. Process.*, vol. 63, no. April 2020, pp. 24–34, 2021, doi: 10.1016/j.jmapro.2020.03.018.
4. J. Xiong, Y. Lei, H. Chen, and G. Zhang, "Fabrication of inclined thin-walled parts in multi-layer single-pass GMAW-based additive manufacturing with flat position deposition," *J. Mater. Process. Technol.*, vol. 240, pp. 397–403, 2017, doi: 10.1016/j.jmatprotec.2016.10.019.
5. L. Yuan et al., "Investigation of humping phenomenon for the multi-directional robotic wire and arc additive manufacturing," *Robot. Comput. Integr. Manuf.*, vol. 63, no. July 2019, p. 101916, 2020, doi: 10.1016/j.rcim.2019.101916.
6. U. Gratzke, P. D. Kapadia, J. Dowden, J. Kroos, and G. Simon, "Theoretical Approach to The Humping Phenomenon in Welding Processes," *J. Phys. D. Appl. Phys.*, vol. 25, no. 11, pp. 1640–1647, 1992, doi: 10.1088/0022–3727/25/11/012.
7. E. Soderstrom and P. Mendez, "Humping mechanisms present in high speed welding," *Sci. Technol. Weld. Join.*, vol. 11, no. 5, pp. 572–579, 2006, doi: 10.1179/174329306X120787.
8. T. C. Nguyen, D. C. Weckman, D. A. Johnson, and H. W. Kerr, "High speed fusion weld bead defects," *Sci. Technol. Weld. Join.*, vol. 11, no. 6, pp. 618–633, 2006, doi: 10.1179/174329306X128464.
9. H. W. Choi, D. F. Farson, and M. H. Cho, "Using a Hybrid Laser Plus GMAW Process," *Weld. J.*, no. August, pp. 174–179, 2006.
10. P. Kazanas, P. Deherkar, P. Almeida, H. Lockett, and S. Williams, "Fabrication of geometrical features using wire and arc additive manufacture," *Proc. Inst. Mech. Eng. 3Part B J. Eng. Manuf.*, vol. 226, no. 6, pp. 1042–1051, 2012, doi: 10.1177/0954405412437126.
11. L. Wang, C. S. Wu, and J. Q. Gao, "Suppression of humping bead in high speed GMAW with external magnetic field," *Sci. Technol. Weld. Join.*, vol. 21, no. 2, pp. 131–139, 2016, doi: 10.1179/1362171815Y.0000000074.
12. L. Yuan et al., "Investigation of humping phenomenon for the multi-directional robotic wire and arc additive manufacturing," *Robot. Comput. Integr. Manuf.*, vol. 63, no. November 2019, p. 101916, 2020, doi: 10.1016/j.rcim.2019.101916.

Index